HEALTH AND MEDICINE
PROJECTS FOR YOUNG SCIENTISTS

KAREN E. O'NEIL

HEALTH AND MEDICINE PROJECTS FOR YOUNG SCIENTISTS

PROJECTS FOR YOUNG SCIENTISTS

FRANKLIN WATTS
NEW YORK | CHICAGO | LONDON | TORONTO | SYDNEY

The Astrand Bicycle Ergometer Test that appears in *Textbook of Work Physiology* by P.O. Astrand, copyright © 1986, is reproduced in this book with permission of McGraw-Hill, Inc.

Photographs copyright ©: Henry Rasof: pp. 11, 12; Fundamental Photos/ Kip Peticolas: p. 16; Randy Matusow: pp. 35, 109; Photo Researchers Inc.: pp. 61, 84 (both Omikron), 72 top (Walter Dawn), 94 (David M. Phillips/ Population Council), 98 (Eric V. Grave); Pfizer Inc. Photo Library: p. 72 bottom.

Library of Congress Cataloging-in-Publication Data

O'Neil, Karen E.
    Health and medicine projects for young scientists / Karen E. O'Neil.
        p.   cm. — (Projects for young scientists)
    Includes bibliographical references and index.
    Summary: Provides instructions for experiments demonstrating medical principles and explains how similar investigations have helped find cures or treatments for serious illnesses.
    ISBN 0-531-11050-8 (lib. bdg.)—ISBN 0-531-15668-0 (pbk.)
    1. Medicine—Experiments. 2. Health—Experiments. 3. Science projects.   [1. Medicine—Experiments. 2. Health—Experiments. 3. Experiments. 4. Science projects.]   I. Title. II. Series.
R852.054 1993
610'.78—dc20   92-42745 CIP AC

CONTENTS

For my parents, Isobel and Edward Thode,
who taught me the value of questioning

1

HEALTH AND MEDICINE PROJECTS

Why does the blood of two patients sometimes clump when mixed? The physician Karl Landsteiner asked this very question and proposed the existence of different blood types to answer it. This led to the first widespread successful use of blood transfusions.

What causes beriberi, a disease whose symptoms include weakness, anemia, and paralysis? Christiaan Eijkman was a physician sent to the Dutch East Indies to investigate this disease. Most of the chickens he took along as test animals died of symptoms similar to those of beriberi. No organism could be found that caused the disease. Why did the chickens die? His investigation finally revealed that the chickens had been fed a diet high in polished rice, leading to the discovery that polishing the rice removes the B vitamins necessary to prevent beriberi.

Do diseases like cancer result from viruses, heredity, diet, environmental factors, all of the above, or something else? Thousands of scientists around the world are tackling the question and, of late, making some progress.

"Doing" science or medicine is a bit like being a detective. In fact, some of the best detectives of all time have been scientists—including those described here—trying to solve medical mysteries. In both detective work and science and medicine, there's a problem or puzzle, there are clues, there are ways to go about trying to solve the problem, and there's often a solution. Scientists, like detectives, believe that almost every problem has a solution—if only enough evidence or information can be obtained and analyzed correctly. However, in detective work as in science and medicine, the solutions to some problems are stubborn or unsolvable at a particular time in history.

Although scientists and medical researchers understand a great deal about health and medicine, the opportunities for research are virtually limitless. While totally original ideas can be hard to tackle without professional training, a lab, assistants, and money, there are still many new areas you can investigate.

WHAT IS A HEALTH PROJECT OR MEDICINE PROJECT AND WHY DO ONE?

Projects in the field of health and medicine are particularly interesting because they investigate questions about the human body: How does it work? What does it need to stay free of disease? How do we learn?

Medical researchers are scientists who attempt to answer questions such as these. Their findings are often used by health care professionals—nurses, lab technologists, and physicians—in the practice of medicine.

Photos 1 and 2. Two health and medicine projects prepared for the Boulder Valley School District Regional Science Fair, in Boulder, Colorado

Sleep Learning-Does it work?

COLLECTED DATA

PURPOSE

OBSERVATIONS AND RESULTS

HYPOTHESIS

PROCEDURE

CONCLUSION

ARE YOU STRESSED YET?

Is the Perception of the Autonomic Nervous System Response Based on Context?

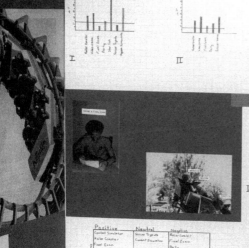

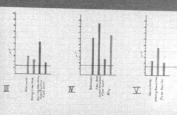

LEGEND
Positive—
Negative—
Neutral—

DATA

RESULTS

CONCL

Positive	Neutral	Negative
Combat Simulator	Soccer Tryouts	Roller Coaster
Roller Coaster	Combat Simulator	Final Exam
Final Exam		Party
Soccer Game		Watching Broncos Game
Party		Mystery Removed
Watching Star Trek		Boring Bike Game
Pain Neutral		
Excited		
Hot and Tired		Hot and Tired

MATERIALS

GY

Photo 3. Another project at the
Boulder science fair

When you do a health or medicine project, you become a temporary medical researcher. The project you choose may be a library investigation or a short-term experiment to fulfill a class assignment. The projects suggested in this book also make good science fair entries, and because the projects are about humans, they give you an opportunity to learn more about yourself (see photos 1–3).

CHOOSING A TOPIC TO INVESTIGATE

If you have never done a science experiment or project on your own before, the hardest part may be getting started. That means identifying a topic or question you'd like to investigate. When faced with a whole world of possibilities, thinking of one question you might be able to answer seems like a huge task.

It doesn't have to be. Consider your interests; eating, fitness, training, stress reduction, high technology. Your interests are a great starting point because they help ensure that you'll stick with your project. Once you have identified your interests, you need to narrow the field and focus on something specific.

Reading through this book will help you focus on a definite question or idea. For example, your major interests may initially be health and sports. This is a big topic, too big for any one person to research. For help in identifying a specific project or question to investigate, go to Chapter 2. Your interests might draw you to the subheading "Training Effects," where you can choose from a number of ideas.

Some of the ideas given in this book are accompanied by information on how to proceed and what kinds of materials and equipment you may need. Others are accompanied by suggested ways to approach them. Still others challenge you to figure out your own path to a solution. There are numerous possibilities for experiments and projects at all levels of ability.

Although coming up with a totally new research idea is not so easily done, there is great value in trying "tried-and-true" experiments and projects for yourself, and of course you may be able to create—with some supervision—your own variations. But each project, whatever the level, encourages imagination and creativity.

Ideas are all around you. Just *thinking* about what to do can be frustrating. Perhaps one of these projects will steer you toward your own idea. Or if nothing here interests you, seek elsewhere. Reading, searching, and listening *while* you're thinking are much more productive than thinking alone.

SCIENCE FAIR COMPETITIONS

Most state and national science fair competitions have deadlines in early to late winter. Annual competitions on the national level include the Duracell/NSTA, Thomas Edison/Max McGraw, International Science and Engineering, Student Space Involvement, and Westinghouse competitions. Information is generally sent to high school and middle school science teachers each fall. You can also find advertisements in *The Science Teacher* and *The American Biology Teacher* magazines. The September issues usually contain the most information.

If the project you're doing will be entered into a science fair competition, you will need to keep in mind the rules of that competition. It is best to check with your teacher or science fair adviser at each step of the process when there are specific guidelines. The International Science and Engineering Fair (ISEF) guidelines require that a Research Plan be approved by the school's science fair committee before the beginning of certain kinds of projects. If your local fair is affiliated with ISEF, you will be required to follow their rules. To be safe, always have your experimental plans approved *before* beginning your investigation.

SAFETY

A preapproved research plan is essential to success for student or professional scientists. Minimizing or eliminating safety hazards is also essential. Areas of special safety concern are emphasized throughout this book, but they are *not* a substitute for a safety check of *your* plan. Make sure you know the proper safety procedures for every project you work on. Discuss those procedures with your teacher before beginning. Your teacher may decide that the project is too dangerous for you to try, for example, because the necessary safety equipment is unavailable. In that case, find another project to do.

In addition, you should learn certain rules for working safely in a science laboratory. Here are some of those rules:

1. Always work under the supervision of a knowledgeable adult such as a qualified science teacher or scientist. When working with chemicals or flame or in any potentially hazardous situation, be sure a science teacher or other professional is present. Professional scientists don't work alone. Neither should you.

2. Always wear approved safety goggles when working with chemicals or chemical equipment. Photo 4 shows a common type of safety goggles recommended for use in science labs. Since you never know when something will go wrong, you do not have time to run and get your eye protection if and when you actually need it. Ask your teacher to inspect and approve the goggles you use. Avoid wearing contact lenses when working with chemicals.

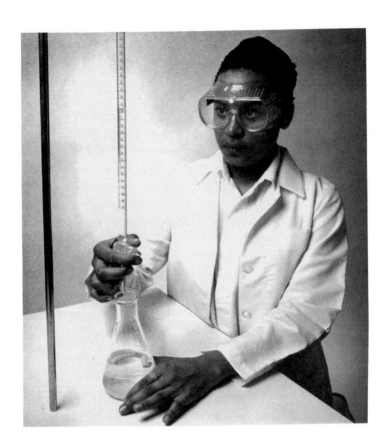

Photo 4. Always wear approved safety goggles and a lab apron or lab coat when working in the lab.

3. Always wear a lab coat, lab apron, or other protective clothing when working with chemicals. Spilling acid on a new shirt is not the tragedy that getting acid in your eye would be, but why lose the shirt? A lab coat or apron, such as the one shown in Photo 4, will help protect you from this kind of accident.

4. Always assume that a chemical or microorganism is dangerous unless you absolutely know differently. This means:

a. *Never taste anything, chemical or biological, in a lab.*

b. *Never inhale deeply near an open chemical container or culture of microorganisms.*

c. *Treat all chemicals and microorganisms as if they are capable of damaging your skin or clothing or doing other harm.* This means:

1. Keep *chemicals off your skin.*

2. *Wipe up spilled materials immediately.*

3. *Do not eat or drink while you are doing experiments.*

4. *Never touch a chemical with your hands.* Always use a spatula or some other tool to pick it up. If a chemical touches your skin, wash immediately with soap and water.

5. Become familiar with the safety equipment in your laboratory and know how to implement first-aid and emergency procedures. Ask your teacher where the eye shower, safety shower, fire extinguisher, and fire blanket are located and how they are used. Know how to operate such equipment and have the assurance of your teacher that the items work properly. Make sure the safety equipment is immediately accessible to your work space.

6. Do not dispose of chemical or biological materials down the drain or in the trash. Ask your teacher how these wastes are handled at your school. Also ask about disposal of other materials and cleaning of equipment.

7. Wash your hands after you finish conducting an experiment.

8. Always keep your work area neat and clean, as shown in the figures. Messiness in the work space can lead to serious accidents.

USING ANIMAL SUBJECTS

The use of animals in student research is generally accepted when gaining the knowledge without using animals would be difficult. When you undertake research involving animals, you have the responsibility to care for and maintain them in as comfortable and humane a manner as possible. The use of *invertebrates* (animals without backbones, like insects and shellfish) is encouraged in student experiments. Their small size, relative ease in care, and great variety make them good subjects for research.

On the other hand, the use of *vertebrates* (animals with backbones, like mammals) in experiments is often discouraged and sometimes strictly prohibited, depending on the state or the science fair competition involved. Vertebrate animal experiments can usually be undertaken only in cooperation with and under the direct supervision of science or medical professionals. Many fairs require a review committee to approve such experiments and sometimes *any* experiments, regardless of whether vertebrates are involved.

Whatever the guidelines of your state or science fair, becoming familiar with your experimental animals, their care, and their needs will benefit both you and your project. The National Science Teachers Association has issued guidelines for appropriate use of animals in schools. A copy may be obtained by writing to them at NSTA Special Publications, 3140 N. Washington Blvd., Arlington, VA 22201. A useful article is "Animal Care Use Committees," published in the February 1992 issue of *The Science Teacher,* a magazine your science teacher may subscribe to.

USING HUMAN SUBJECTS

Many intriguing health and medicine questions can be answered only with the participation of human subjects. However, it is essential to prevent any form of physical or psychological risk to humans in a science project.

The federal government has strict regulations concerning the use of humans in research. If your project involves humans (including you) in any way—experimentation, surveys, questionnaires, or data collection—you must get approval of your plan either from your school's Institutional Review Board (part of the science fair structure) or from two people (a science teacher and one of the following: a nurse, physician, or social scientists. Ask a science teacher to help you go through the proper procedures, keeping in mind that a *different* science teacher needs to approve the final plan.

A research plan involving humans often includes a written statement about the procedures that will be used and any possible risks, if they exist. The written statement is generally followed by a space for the subjects (and their parents, if they are less than 18 years old) to sign acknowledging that they understand the information and agree to participate. Proceed with any experiment involving humans only after you have first received written approval from your Institutional Review Board, described in the previous paragraph, and from the subjects of your experiment.

Many exciting projects with humans are possible despite the restrictions. This book includes quite a few suggestions for projects that involve volunteers in both experimental and survey capacities. If you do any of these projects, your primary concern must always be the welfare of your experimental subjects.

A STITCH IN TIME SAVES NINE: THE IMPORTANCE OF PLANNING

Most health and medicine investigations take time and planning. Few projects suggested in this book can be completed in a week or less. To give yourself the best opportunity for succeeding, for really learning something new, you will need to become a part-time medical investigator for one to several months. A science project need not take over your life. It just has to become a regular feature of it.

When you know the question you are going to investigate, devise an experimental plan. Make a list of the equipment and materials you'll need. Organize a space at school or at home (if safety considerations allow it) where you will carry out the work. If materials need to be ordered, find out about order and delivery times. When purchasing and/or using live materials, complete all preparations for care and maintenance of the organisms *before* they arrive.

If you are collecting invertebrates or other specimens from the local area, determine whether they can be found at the time of year you'll be doing your work. Avoid the most common complaint of kids doing science projects— too little time—by starting early.

It is always wise to build in some "disaster" time. The unexpected can happen: You may run out of a chemical before your experiment is finished. You may need to perfect a new technique you're learning. Volunteers may drop out of your study. At worst, you may have to start over. At best, you'll have plenty of time to analyze your results or to pursue a new question which grew out of your work.

MAKING YOUR HEALTH OR MEDICINE PROJECT SCIENTIFIC

There is really no substitute for the excitement of discovering something on your own. Curiosity, interest, and the desire for a new experience are all good reasons to do a science project. But how will you know whether your findings mean anything?

At the outset, you will want to be sure that your experiment is well designed. Some features of well-designed experiments include limitation of variables, inclusion of a control, elimination of bias, and use of sufficient numbers of tests.

Limiting variables simply means looking at one question or feature of a situation at one time. For instance, you may wish to investigate the effect of antibiotics on bacterial growth. If three antibiotics are mixed and placed in a culture, you will not know whether one or two or all three together have had an effect on the bacteria. More useful information would be produced by testing each antibiotic separately for its effect. *Including a control* is necessary for comparison purposes. If an antibiotic is placed in a bacterial culture and the bacteria fail to grow, you cannot conclude that the lack of growth was caused by the antibiotic *unless* you can show that the bacteria would grow under the same conditions *without* the antibiotic. A *control* is an untreated comparison group which allows you to make conclusions.

Bias is a predisposition to a certain result. There are two kinds of bias to avoid in experimentation. One is *investigator bias,* when the person conducting the experiment influences the results because he or she expects a certain outcome. To minimize or prevent investigator bias, include numerical measurements, called quantitative data, in your experimental design as well as descriptive, or qualitative, data. If you are using human subjects, do not let them know whether they are in the control or the experimental treatment group, if possible. This prevents the subjects' giving you what they think you expect.

The second type of bias applies to surveys and is called *selection bias.* The best way to ensure that two groups are as similar as possible is randomly to assign volunteers to one group, the control, or the other, the experimental. This is called randomized, controlled experimentation. In conducting a survey, attempt random sampling where possible. This means using a chance process to select survey participants from your identified population.

Finally, the validity of your results often rests with the *amount of data* you have collected. How many tests, subjects, or animals are needed for an experiment? That, in part, depends on the outcome. But since you can't know that ahead of time, do as many tests as are reasonably practical for your time, space, and expense limitations. Remember, larger numbers (scientists may test hundreds or thousands of subjects, for example) minimize the possibility of chance differences cropping up in your results.

It is essential to keep good records of what you have done and how you have done it. Keep these data, both quantitative (involving precise measurement) and qualitative (involving more subjective descriptions) types of observations, in a special notebook. Get in the habit of making daily dated entries in the notebook during your project. What may seem unimportant at the moment could turn out to be an essential observation at a later date. Know what you found out and when.

GETTING HELP

Don't be reluctant to ask for help in locating specialized equipment, answering a planning question, or learning a new technique. Your science teacher or science fair adviser should be able to direct you to an adult or other student who can help. Although many students do successful projects with minimal guidance, many or most of these who go on to win top honors in national science fair competitions work with experienced science teachers or professional scientists.

DON'T STOP BEFORE THE FINISH LINE

The results are in. The experimenting has been finished. What does it all mean? Until you analyze your data, you don't know what you've accomplished.

Is a difference you have found due to chance or something else? Only *tests of significance* can answer this question. Applying a test of significance need not be difficult, yet it can take much of the guesswork out of drawing conclusions. Why is this the part of a science project most people skip? How can you avoid this trap? You have done the creative work; you've completed the experimenting; don't stop now!

Find a book on statistics (several are suggested at the end of this chapter). Read about tests of significance. Most likely the *z test* or the *chi-square test* will apply to your results if you are working from a hypothesis or have done a controlled experiment where you were looking for differences. Seek the advice and aid of a math or science teacher. All of this should help you find out what you've discovered in your experiment!

PRESENTING YOUR PROJECT

If you are entering your project in a science fair, be careful to follow their guidelines in constructing your display. Generally speaking, a sturdy, uncluttered display board is most effective.

Convey your project title or question, and present the basic elements of your project. Make both your plan and your results clear to the observer. A few well-chosen photographs or carefully constructed graphs can enhance your data section and draw interest. Don't forget to have your original data books as well as the final report available for inspection.

THE END IS IN SIGHT

Albert Einstein is credited with saying, "No amount of experimentation can ever prove me right; a single experiment can prove me wrong." He has captured the essence of science very simply. In science you have to be ready to be wrong in your predictions. It doesn't matter that things turned out differently than you expected; you have discovered something on your own. When you have your results, being *right* isn't important. Showing that something *isn't* is as important as showing that something *is.*

You have reached the end, the point at which you are confident you have interpreted your results correctly. Are you really finished? Isn't there something else you'd like to investigate now? What new questions has your work raised? Another question, another search. Isn't this what science really is?

READ MORE ABOUT IT

Jaeger, Richard M. *Statistics: A Spectator Sport.* Beverly Hills, Calif.: Sage, 1983.

Kotz, Samuel, *Educated Guessing.* New York: Marcel Dekker and D. F. Stroup, 1983.

Moore, David. *Statistics, Concepts and Controversies.* San Francisco: W. H. Freeman, 1979.

Riedel, Manfred G. *Winning with Numbers: A Kids' Guide to Statistics.* Englewood Cliffs, N.J.: Prentice-Hill, 1978.

Rowantree, Derek. *Statistics without Tears.* New York: Scribner's, 1981.

Seuling, Barbara. *You Can't Sneeze with Your Eyes Open and Other Freaky Facts about the Human Body.* New York: Dutton, 1986.

Smith, Norman. *How to Do Successful Science Projects.* Englewood Cliffs, N.J.: Julian Messner, 1990.

Thomas Lewis. *The Youngest Science: Notes of a Medicine Watcher.* New York: Bantam, 1984.

Tocci, Salvatore. *How to Do a Science Fair Project.* New York: Franklin Watts, 1986.

Van Deman, Barry A., and E. McDonald. *Nuts and Bolts: A Matter of Fact Guide to Science Fair Projects.* Harwood Heights, Ill.: The Science Man Press, 1980.

2

FITNESS, SPORTS, AND TRAINING

Most human beings have the potential to become physically fit. The quality of our lives, our energy level, and our health are all improved when we are physically fit. Heredity, environment, health history, and other factors determine the limits of our physical abilities. However, each of us has a choice, within those limits, of whether to lead a healthy or unhealthy physical life-style. While few of us ever explore the upper limits of our physical abilities, we are able to remain fit and reap the health benefits of physical fitness.

Physical fitness today is defined more broadly than it was in the past. Important components include heart-lung ("aerobic") fitness, relative leanness, flexibility, muscular strength, and endurance. Within each of these areas, numerous tests have been designed to evaluate fitness levels. The information is often used to set personal goals for improvement or, more recently, to find out which sport or sports a person is most suited to.

If you are interested in sports and fitness, there are many questions for you to investigate. The activities suggested in this chapter often involve human subjects and require the cooperation of a physical education professional, coach, or fitness instructor. You will need to review the guidelines in Chapter 1 and plan ahead in order to obtain meaningful information.

Guidelines for Working with Human Subjects: All projects in this chapter should be done with the supervision of a qualified science teacher and health care professional. All subjects, including you, should be screened by a physician or nurse and should consult an exercise technician—or all three. See Chapter 1 for more information on the use of human subjects.

FITNESS MEASUREMENTS

Limits to physical activity are of two basic types; muscle strength and energy. If you have to stop an activity, it's usually because your muscle fatigues or you get out of breath. For aerobic activities such as running, swimming, or cycling, the fitness of the cardiorespiratory, or heart-lung, system determines the amount of oxygen delivered to the muscles. Without oxygen, energy cannot be efficiently provided to the muscles for their continued use.

Several variables can be used to help evaluate cardiorespiratory fitness, among them heart rate, blood pressure, lung capacity, and oxygen uptake.

TESTING ENDURANCE BY MEASURING VO$_2$MAX

A frequently used measurement of fitness and potential for endurance sports is the VO$_2$max. The VO$_2$max is the highest amount of oxygen that can be used by the body during hard work. Competitive marathoners, cross-country skiers, and cyclists have high VO$_2$max measurements. Their bodies are able to use a great deal of oxygen for muscular work.

Are you curious about your cardiorespiratory fitness? Can training for an endurance activity, like distance running or swimming, increase your VO_2max?

To investigate this question, you will need to test yourself before beginning a training program and again after several weeks of training. The only equipment you need is an exercise bicycle.

First, ask a physical education instructor, coach, or fitness instructor to help you design an endurance training program. Such training is usually meant to exercise the heart and lungs vigorously for at least 20 minutes at a time.

Next, get your physician's okay to proceed with the plan. This precaution is always suggested whenever a person begins a new or demanding exercise program, in the event that unknown cardiac, respiratory, or other problems are present.

You now need to determine VO_2max, which you will do by taking a ride on your stationary bike.

Start the test at least 2 hours after a meal and when you feel well rested. Adjust the bicycle seat and handlebars so that your knee is slightly bent at the bottom position. Refer to Figure 1 to check the proper position of the seat relative to your feet on the pedals. A knee angle of 160 degrees is suggested and can be measured using the goniometer illustrated in Figure 2 and described in a later project. You should be comfortable, with handlebar and seat placement positions noted for replication on any future tests.

Begin by pedaling at a rate of 60 revolutions per minute (RPM) for a few minutes. This is your warm-up, and it is *essential* as a health precaution and for the success of the experiment. Then adjust the load, the amount of friction on the wheel that increases pedaling resistance. Turn the load up to the highest point where you feel you could

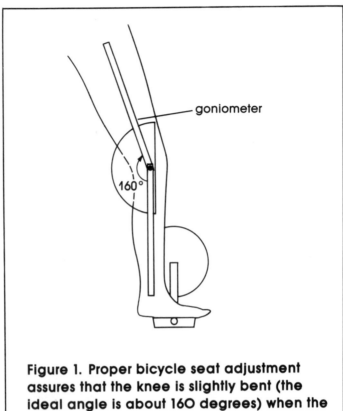

goniometer

160°

Figure 1. Proper bicycle seat adjustment assures that the knee is slightly bent (the ideal angle is about 160 degrees) when the pedal is in the lowest position.

pedal comfortably for 10 minutes. After pedaling at this setting for 4 minutes, start timing the test. Pedal for the remaining 6 minutes and take your pulse during the last minute of the test. Cool down for several minutes by pedaling at a slower rate. *Never quit vigorous exercise suddenly.*

Use your pulse (heart rate) and workload (the power setting on the bicycle) to find the largest amount of oxygen you consume (your maximal oxygen uptake) during

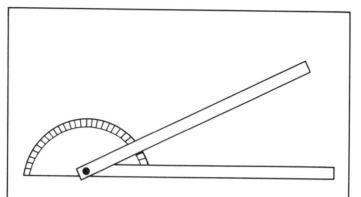

Figure 2. This goniometer, made from a protractor and two straightedges, can be used to measure the maximum angle in a joint or the range of motion around a joint.

this test in Table 1 or Table 2. Workload is usually measured in watts (W), a unit of power, or in kilopons-meters (KPM), a measure of resistance. Next go to Table 3 to determine your VO_2max. For example, according to Table 1, at a heart rate of 120 and a workload of 300 KPM/min (50 W), a male would have a maximal oxygen uptake of 2.2 liters per minute. According to Table 3, the person, if he weighed 140 pounds, would have a VO_2max on maximal oxygen uptake of 34 liters per minute. Compare pretraining and posttraining VO_2max values to determine whether they have been affected by the training.

The results you obtain for yourself may not be applicable to others. If you wish to attempt a broader look at questions involving VO_2max, here are some additional ideas for projects. Be sure to screen all subjects for potential health problems.

TABLE 1. Men: Maximal Oxygen Uptake from Heart Rate and Work Load

ASTRAND BICYCLE ERGOMETER TEST

Heart rate	Maximal oxygen uptake (liters per minute)				
	300 KPM/min 50W	600 KPM/min 100W	900 KPM/min 150W	1,200 KPM/min 200W	1,500 KPM/min 250W
120	2.2	3.5	4.8		
121	2.2	3.4	4.7		
122	2.2	3.4	4.6		
123	2.1	3.4	4.6		
124	2.1	3.3	4.5	6.0	
125	2.0	3.2	4.4	5.9	
126	2.0	3.2	4.4	5.8	
127	2.0	3.1	4.3	5.7	
128	2.0	3.1	4.2	5.6	
129	1.9	3.0	4.2	5.6	
130	1.9	3.0	4.1	5.5	
131	1.9	2.9	4.0	5.4	
132	1.8	2.9	4.0	5.3	
133	1.8	2.8	3.9	5.3	
134	1.8	2.8	3.9	5.2	
135	1.7	2.8	3.8	5.1	
136	1.7	2.7	3.8	5.0	
137	1.7	2.7	3.7	5.0	
138	1.6	2.7	3.7	4.9	
139	1.6	2.6	3.6	4.8	
140	1.6	2.6	3.6	4.8	6.0
141		2.6	3.5	4.7	5.9
142		2.5	3.5	4.6	5.8
143		2.5	3.4	4.6	5.7
144		2.5	3.4	4.5	5.7
145		2.4	3.4	4.5	5.6
146		2.4	3.3	4.4	5.6
147		2.4	3.3	4.4	5.5
148		2.4	3.2	4.3	5.4
149		2.3	3.2	4.3	5.4
150		2.3	3.2	4.2	5.3
151		2.3	3.1	4.2	5.2
152		2.3	3.1	4.1	5.2
153		2.2	3.0	4.1	5.1
154		2.2	3.0	4.0	5.1
155		2.2	3.0	4.0	5.0
156		2.2	2.9	4.0	5.0
157		2.1	2.9	3.9	4.9
158		2.1	2.9	3.9	4.9
159		2.1	2.8	3.8	4.8
160		2.1	2.8	3.8	4.8
161		2.0	2.8	3.7	4.7
162		2.0	2.8	3.7	4.6
163		2.0	2.8	3.7	4.6
164		2.0	2.7	3.6	4.5
165		2.0	2.7	3.6	4.5
166		1.9	2.7	3.6	4.5
167		1.9	2.6	3.5	4.4
168		1.9	2.6	3.5	4.4
169		1.9	2.6	3.5	4.3
170		1.8	2.6	3.4	4.3

TABLE 2. Women: Maximal Oxygen Uptake from Heart Rate and Work Load

ASTRAND BICYCLE ERGOMETER TEST

Maximal oxygen uptake (liters per minute)

Heart rate	300 KPM/min 50W	450 KPM/min 75W	600 KPM/min 100W	750 KPM/min 125W	900 KPM/min 150W
120	2.6	3.4	4.1	4.8	
121	2.5	3.3	4.0	4.8	
122	2.5	3.2	3.9	4.7	
123	2.4	3.1	3.9	4.6	
124	2.4	3.1	3.8	4.5	
125	2.3	3.0	3.7	4.4	
126	2.3	3.0	3.6	4.3	
127	2.2	2.9	3.5	4.2	
128	2.2	2.8	3.5	4.2	4.8
129	2.2	2.8	3.4	4.1	4.8
130	2.1	2.7	3.4	4.0	4.7
131	2.1	2.7	3.4	4.0	4.6
132	2.0	2.7	3.3	3.9	4.5
133	2.0	2.6	3.2	3.8	4.4
134	2.0	2.6	3.2	3.8	4.4
135	2.0	2.6	3.1	3.7	4.3
136	1.9	2.5	3.1	3.6	4.2
137	1.9	2.5	3.0	3.6	4.2
138	1.8	2.4	3.0	3.5	4.1
139	1.8	2.4	2.9	3.5	4.0
140	1.8	2.4	2.8	3.4	4.0
141	1.8	2.3	2.8	3.4	3.9
142	1.7	2.3	2.8	3.3	3.9
143	1.7	2.2	2.7	3.3	3.8
144	1.7	2.2	2.7	3.2	3.8
145	1.6	2.2	2.7	3.2	3.7
146	1.6	2.2	2.6	3.2	3.7
147	1.6	2.1	2.6	3.1	3.6
148	1.6	2.1	2.6	3.1	3.6
149		2.1	2.6	3.0	3.5
150		2.0	2.5	3.0	3.5
151		2.0	2.5	3.0	3.4
152		2.0	2.5	2.9	3.4
153		2.0	2.4	2.9	3.3
154		2.0	2.4	2.8	3.3
155		1.9	2.4	2.8	3.2
156		1.9	2.3	2.8	3.2
157		1.9	2.3	2.7	3.2
158		1.8	2.3	2.7	3.1
159		1.8	2.2	2.7	3.1
160		1.8	2.2	2.6	3.0
161		1.8	2.2	2.6	3.0
162		1.8	2.2	2.6	3.0
163		1.7	2.2	2.6	2.9
164		1.7	2.1	2.5	2.9
165		1.7	2.1	2.5	2.9
166		1.7	2.1	2.5	2.8
167		1.6	2.1	2.4	2.8
168		1.6	2.0	2.4	2.8
169		1.6	2.0	2.4	2.8
170		1.6	2.0	2.4	2.7

TABLE 3. VO$_{2MAX}$ FROM MAXIMAL OXYGEN UPTAKE

ASTRAND BICYCLE ERGOMETER TEST

Maximal oxygen uptake (liters per minute) vs. Weight (pounds)

L/min	80	85	90	95	100	105	110	115	120	125	130	135	140	145	150	155	160	165	170	175	180	185	190	195	200	205	210	215	220	225	230	235	240	245	250	255	260
2.0	56	51	49	47	43	42	40	38	36	35	34	33	31	30	29	29	27	27	26	25	24	24	23	22	22	22	21	20	20	20	19	19	18	18	18	17	17
2.2	61	56	54	51	48	46	44	42	40	39	37	36	34	33	32	31	30	29	29	28	27	26	26	25	24	24	23	22	22	22	21	21	20	20	19	19	19
2.4	67	62	59	56	52	50	48	46	44	42	41	39	38	36	35	34	33	32	31	30	29	29	28	27	26	26	25	24	24	24	23	22	22	22	21	21	20
2.6	72	67	63	60	57	54	52	50	47	46	44	43	41	39	38	37	36	35	34	33	32	31	30	29	29	28	27	27	26	25	25	24	24	23	23	22	22
2.8	78	72	68	65	61	58	56	54	51	49	47	46	44	42	41	40	38	37	36	35	34	33	33	31	31	30	29	29	28	27	27	26	26	25	25	24	24
3.0	83	77	73	70	65	63	60	58	55	53	51	49	47	45	44	43	41	40	39	38	37	36	35	34	33	32	32	31	30	29	29	28	28	27	26	26	25
3.2		82	78	74	70	67	64	62	58	56	54	52	50	48	47	46	44	43	42	40	39	38	37	36	35	34	34	33	32	31	30	30	29	29	28	28	27
3.4			83	79	74	71	68	65	62	60	58	56	53	52	50	49	47	45	44	43	41	40	40	38	37	37	36	35	34	33	32	32	31	31	30	29	29
3.6				84	78	75	72	69	65	63	61	59	56	55	53	51	49	48	47	45	44	43	42	40	40	39	38	37	36	35	34	34	33	32	32	31	31
3.8					83	79	76	73	69	67	64	62	59	58	57	54	52	51	49	48	46	45	44	43	42	41	40	39	38	37	36	36	35	34	33	33	32
4.0						83	80	77	73	70	68	66	63	61	59	57	55	53	52	50	49	48	47	45	44	43	42	41	40	39	38	37	37	36	35	34	34
4.2							84	81	76	74	71	69	66	64	62	60	58	56	55	53	51	50	49	47	46	45	44	43	42	41	40	39	39	38	37	36	36
4.4								85	80	77	75	72	69	67	65	63	60	59	57	55	54	52	51	49	48	47	46	45	44	43	42	41	40	40	39	38	37
4.6									84	81	78	75	72	70	68	66	63	61	60	58	56	55	53	52	51	49	48	47	46	45	44	43	42	41	40	40	39
4.8										84	81	79	75	73	71	69	66	64	62	60	59	57	56	54	53	52	51	49	48	47	46	45	44	43	42	41	41
5.0											85	82	78	76	74	71	68	67	65	63	61	60	58	56	55	54	53	51	50	49	48	47	46	45	44	43	42
5.2												85	81	79	76	74	71	69	68	65	63	62	60	58	57	56	55	53	52	51	50	49	48	47	46	45	44
5.4													84	82	79	77	74	72	70	68	66	64	63	61	59	58	57	55	54	53	51	50	50	49	47	47	46
5.6														85	82	80	77	75	73	70	68	67	65	63	62	60	59	57	56	55	53	52	51	50	49	48	47
5.8															85	83	79	77	75	73	71	69	67	65	64	62	61	59	58	57	55	54	53	52	51	50	49
6.0																86	82	80	78	75	73	71	70	67	66	65	63	61	60	59	57	56	55	54	53	52	51
6.2																	85	83	81	78	76	74	72	70	68	67	65	63	62	61	59	58	57	56	54	53	53

- Do members of endurance sports teams, such as cross-country, show an increase in VO_2max during the season?
- If VO_2max does increase with training, what types of individuals see the greatest increase on a percentage basis, novices or returning team members?
- Are there gender differences in VO_2max for individuals of the same weight?
- If VO_2max does increase with training, how long does it take to see an effect?
- Does the *type* of training regimen influence VO_2max changes?
- Is there a difference between mean VO_2max values of individuals who are subjected to passive smoke in their homes and those who are not?

OTHER ENDURANCE INDICATORS

Oxygen use is a good indicator of cardiorespiratory fitness, but indirect physiological indicators of oxygen use by the muscles include heart rate, blood pressure, and pulmonary ventilation rate (the amount of air exhaled in 1 minute). Do these physiological indicators decrease or increase when endurance training is undertaken?

Monitor these factors in yourself first, before and after you undertake an endurance training program. Be sure to work under supervision. Then you can also:

- Compare preseason and postseason resting rates of the members of various sports teams. Do any of the teams show a change in mean resting heart rate?
- Find out whether the pulmonary ventilation rates of athletes and nonathletes are different. Use a respirometer, available in many high school sci-

ence labs, or a large balloon in an experiment of your own design to answer this question. Be careful to account for gender and weight variables.

- Look into whether the amount of playing time within a sports team produces a training effect on VO_2max, heart rate, blood pressure, or pulmonary capacity during the sports season.

OTHER ENDURANCE TESTS

- Investigate a test called the Cooper 12-minute run, developed by Kenneth Cooper, M.D. He has written several books that provide information on this test.
- Investigate the Harvard step test. A good reference by Kirkendall is listed at the end of this chapter.

TRAINING EFFECTS: MUSCLE STRENGTH AND ENDURANCE

Does weight training really build muscles? Muscle strength—the amount of force exerted by muscles—and endurance—the amount of time muscles can exert force—allow people to carry out daily activities without becoming exhausted. You may be curious about the effectiveness of various weight training regimens. One question about weight training is this: Is there a difference between training at lighter loads with more repetitions and training at heavier loads with fewer repetitions? Can males and females both build big muscles? See Photo 5 for a challenging question.

To do this investigation you will need volunteer subjects of the same sex and access to weight training equipment.

Safety Note: Be sure to work under supervision and screen subjects for potential medical problems. Volunteers under the age of 16 years should lift no more than

Photo 5. Does musculature alone tell you which arm belongs to a female and which to a male? What other clues do you need?

25 to 50 percent of their body weight in a weight training program.

Randomly divide your group of volunteers into two groups. For each, determine the maximum weight the individual can lift and lower three times in succession. This is called the individual's three-repetition maximum, or 3RM. Also determine the individual's 10RM in a similar manner. Have one group do ten sets of 3RM (low repetition/high load) and the other group ten sets of 10RM (high repetition/low load) three times a week, with at least a day of rest between sessions, for 6 to 9 weeks. Adjust individual RM each session as needed. In other words, increase or decrease the load if the subject's strength and endurance change. At the end of the training period, compare strength gains. What do you conclude?

Here are some other, related investigations:

- How many weeks does it take for strength to double in each kind of weight training regimen?
- Is there an optimum number of sets per workout beyond which no further gains in strength are realized?
- Does lower-body training—for example, training of the quadriceps—follow a pattern similar to that in upper-body training?
- Do high load/low repetition and low load/high repetition regimens lead to similar or different changes in muscle girth for males?
- Do females achieve strength or muscle girth increase similar to males' by following the same training regiments?
- Do high repetition/low load regimens result in superior muscle endurance gains when compared with low repetition/high load regimens?
- Does lower-body weight training lead to improved performance in track or cycling events?
- Does upper-body weight training lead to im-

proved performance in swimming and field events?

THE COMPETITIVE EDGE: FLEXIBILITY, AGILITY, AND COORDINATION

Flexibility, or range of motion (ROM) within a joint, as well as agility and coordination all contribute to the harmonious execution of physical tasks that involve many muscles. On the basketball court or the gymnastics mat, these attributes combine to produce both beautiful movement and effective competition.

You may be interested in the contribution of flexibility to performance in physical activities. You can measure range of movement in many joints with a device called a goniometer. It can be made from a protractor and two movable straight edges affixed to the center of the protractor (refer to Figure 2).

- Does the ROM of a pitcher's throwing arm differ from that of baseball players in other positions?
- Does the ROM of a hurdler's leg differ from that of other track athletes?
- In which sports or physical activities do individuals have the highest ROMs for arm, leg, wrist, or ankle joints?
- What kind of training improves ROMs for specific joints?
- Does weight training have an effect on ROM?

AGILITY AND COORDINATION

What abilities combine to produce a successful pole vault, a superbly executed basketball dunk, or a flawlessly executed grand jeté? Do some reading on tests of agility and coordination. Do these tests help predict or determine success in physical activities? Can agility and coordination be improved through training?

READ MORE ABOUT IT

Arnot, Robert Burns, and Charles Gaines. *Sportselection.* New York: Viking, 1984.

Bershad, Carol, and Deborah Bernick. *Bodyworks: The Kids' Guide to Food and Physical Fitness.* New York: Random House. 1979.

Cooper, Kenneth, *Aerobics.* New York: Bantam, 1977.

Fisher, A. Garth, and Clayne R. Jensen. *Scientific Basis of Athletic Conditioning.* Philadelphia: Lee and Febiger, 1990.

Fixx, James F. *Maximum Sports Performance.* New York: Random House, 1985.

Fleck, S. J., and W. J. Kraemer. *Designing Resistance Training Programs.* Champaign, Ill.: Human Kinetics Books, 1987.

Garrick, James G., and Peter Radetsky. *Be Your Own Personal Trainer.* New York: Crown, 1989.

Health Related Physical Fitness Test Manual. Reston, Va. American Alliance for Health, Physical Education, Recreation and Dance, 1980.

Howley, Edward T., and B. Don Franks. *Health and Fitness Instruction.* Champaign, Ill.: Human Kinetics Books, 1986.

Kirkendall, D., J. Gruber, and R. Johnson. *Measurement and Evaluation for Physical Educators.* Dubuque, Iowa: Wm. C. Brown, 1980.

Mangi, Richard, Peter Jokl, and O. William Dayton. *Sports Fitness and Training.* New York: Pantheon, 1987.

Ward, Brain R. *Exercise and Fitness.* New York: Franklin Watts, 1988.

3

DIET AND NUTRITION

Eating should be fun. But can you eat your favorite foods and remain healthy? The general consensus among scientists and nutritionists is that what and how much you eat do affect your health, energy level, sense of well-being, and longevity.

Yet Americans spend millions of dollars annually for various foods and diets in an attempt to find answers to their nutrition questions. And because nutrition is a complex, not fully understood topic, even some doctors, medical researchers, and nutritionists can't provide a simple answer. The result is that separating food fact from food myth is not easy.

However, it is well worth knowing the facts about food:
- How much do people know about good nutrition?
- Do they put this knowledge into practice?

The projects in this chapter may help you find some of the answers.

HOW MUCH DO YOUR CLASSMATES KNOW ABOUT NUTRITION?

What beliefs do your classmates have about topics such as megavitamin therapy, sugar, and diet foods? You can find out by carrying out a survey.

Begin by researching beliefs about nutrition in a recent credible nutrition book. Several are listed at the end of this chapter. You may supplement this information with articles from scientific magazines and journals.

Then design a survey containing clearly correct or incorrect responses. You may wish to focus your survey on a particular area such as diets, vitamin and mineral supplements, fast foods, or advertising claims. Administer the survey to at least fifty people, enough to allow patterns of knowledge to appear, if present. Review the guidelines in Chapter 1 to avoid survey pitfalls and biases. Do your classmates achieve a passing grade for nutritional awareness?

Following are other projects to try.

- Does taking a health course increase a person's nutritional awareness? Carry out a study of two groups to find the answer for your school. To find out how health students' understanding of nutrition changes, work with a health teacher to administer your survey at the start and the finish of the health course or nutrition unit.
- Many parents worry that their children don't have the knowledge to choose a nutritional diet. Compare adult and adolescent nutrition awareness by administering the survey to both groups and evaluating for statistically significant differences. Do adults know more about nutrition than adolescents?
- Who has a greater influence on an adolescent's

nutrition awareness, parents or peers? Survey both groups to find out where there are more similarities in knowledge.

- Does knowledge of good nutritional practices usually translate into wise choices? For example, a student may understand the value of a balanced diet and then choose to skip breakfast and eat a candy bar with a soda for lunch. Formulate a survey which includes both nutritional awareness questions and questions about the respondents' nutritional habits.

VITAMIN C

While food contains hundreds of different substances, only about forty are known to be essential to human health. These forty substances, called nutrients, include water, glucose, amino acids (from protein), fatty acids, vitamins, and minerals.

Among these nutrients, perhaps none has generated more controversy than vitamin C. A continuing claim made for vitamin C is that very large doses can prevent or cure colds, cancer, and a host of other diseases. Dozens of medical studies have failed to support this claim for very large doses of vitamin C.

But the debate about megavitamin doses should not overshadow the need for adequate amounts of the vitamin called the Reference Daily Intake, or RDI. Sixty milligrams, the amount of vitamin C in one orange, does help to promote healing, fight infection, and increase iron absorption in the the body.

What foods contain vitamin C and how much of them do you need to eat each day to consume the RDI? You can use a chemical reaction involving a blue dye called indophenol to answer this question. The procedure for analyzing foods for their vitamin C content is found in many biology and chemistry books. One such reference is *Consumer Chemistry Projects for Young Scientists* by David Newton.

Safety Note: While indophenol has no specific toxicity identified with its use, refer to the safety rules concerning use of chemicals in Chapter 1. Wear safety goggles and a lab apron throughout this procedure.

You can use the indophenol test to answer many more questions about vitamin C:

- Does cooking method affect vitamin C content? For example, does frying affect the vitamin C content of potatoes? Test potatoes from the same bag by preparing them in different ways: boiling, pan frying, deep frying, baking, microwaving, and so on. Compare boiled, steamed, and stir-fried broccoli from the same head of fresh broccoli. Compare your results with those you found for raw vegetables.
- Test the effects of food preparation, too. Compare mashed potatoes with potato chips, shoestrings, and homefries. Test with and without skins.
- Many foods contain no vitamin C in their *natural* state but have it added. Do foods with added vitamin C test the same as foods with *natural* vitamin C? Do foods with added vitamin C contain the amounts they claim on their labels? Do any junk foods contain significant amounts of vitamin C?
- Are fruit juices a good source of vitamin C? If so, which juices are best? Find out what happens to the vitamin C content of these fruit juices over time.
- Test soft drinks and fruit "drinks" for vitamin C content—Hi-C, 7-Up, Slice, Kool-aid, etc. Which

drinks deliver the most vitamin C for the lowest cost?

- Which methods of food storage most effectively preserve vitamin C over time? Compare fresh, canned, frozen irradiated, and freeze-dried portions of the same type of fruit or vegetable or juice. What is the most effective way to store uneaten portions of food in order to preserve vitamin C: plastic wrap, foil, waxed paper, Tupperware, aluminum or tin cans, clear or brown glass jars, covered plastic containers, cardboard or foil-lined cardboard containers, etc.? Do light, temperature, and exposure to the air affect the vitamin C content of stored foods? Does container shape affect vitamin C retention? On the basis of what you learn from your experiments, design the ideal container and storage method for, say, orange juice. Is there anything you could teach commercial producers of these products?

- Why do some animals, such as fish and several of the farm animals, manufacture their own vitamin C in their livers?

OTHER VITAMINS

Testing for other vitamins is more elaborate. Ask your science teacher for assistance.

MINERALS

Minerals, unlike vitamins and most other nutrients, contain no carbon. These inorganic substances are necessary for a number of vital body functions, including muscle contraction, conduction of nerve impulses, formation of blood cells, and regulation of metabolism. They also are necessary for bone and tooth health. Humans need at least six minerals in significant amounts and another fourteen in trace amounts to maintain health.

Mineral deficiencies in the diet have been linked with such conditions as osteoporosis, anemia, diabetes, and atherosclerosis. However, the typical balanced diet contains adequate amounts of minerals with the possible exception of iron and zinc.

Do the diets of adolescents supply them with their mineral requirements? You can find out by doing some research and testing of a group of volunteers from your school.

For more than 40 years, various government and independent research agencies have published recommendations for vitamin and mineral intake. The Recommended Dietary allowance (RDA) was the standard for years. In 1992, it was changed to the RDI, or Reference Daily Intake. Information on the RDIs for twelve minerals was printed in the November 1992 Federal Register and undoubtedly will be reprinted in other government publications and in commercials books. This will give you levels of mineral intake recommended by age and gender. Also locate the U.S. Department of Agriculture (USDA) publications that specify the mineral content of various foods. These are listed at the end of this chapter and are usually available in libraries.

Prepare copes of a form like the one in Table 4. (Please do not write in this book.) Next round up ten to twenty volunteers. Ask each to keep a detailed food diary on the form for at least 2 weeks. The food diary should contain daily entries of all food and liquids eaten and drunk, along with estimates of the amounts consumed.

Here are some tips which may improve the validity of your results. Train your volunteers in proper estimation of size of portions, using the units you see reported in the reference publications. Emphasize the importance of reporting *everything* consumed; use examples of items peo-

TABLE 4. PARTIAL SAMPLE FOOD DIARY

Day #   Gender (male or female)

Enter all foods you eat and drink today, along with estimated amounts. Give brand names where available.

Food name	How much/ how many?	Estimated weight
2% milk	2 cups	
banana	1	
Shredded Wheat (Nabisco)	2 biscuits	47.2 grams
M&M's, plain	2 ounces	57 grams

ple often forget to report, like midmorning snacks, handfuls of grapes, bananas grabbed on the way out of the house, or afternoon cans of soda pop. Have the forms returned without names, identifying the individuals only by gender. This may lessen underreporting of eating behavior the individual considers embarrassing. Finally, assure confidentiality and request that normal eating behavior *not* be changed during the time the food diary is being kept.

Now analyze the food diaries. You will need to calculate daily intake of the various minerals, using information in the food diaries. Select each one of the foods reported in the diary and look up its mineral content. Calculate the amount of each mineral consumed. For example, if 1 cup of skim milk contains 296 grams (g) of calcium, 3 cups contain 888 g.

Next, total the mineral amounts consumed in all of the foods each day. Sometimes you will need to substitute a similar item for a food in the diary since there will not be a listing for every food product in the literature. In some cases you may be able to obtain mineral content information from the makers of specific products, directly from the package, and from fast-food outlet managers. You will need to be thorough during this phase of your research, keeping accurate records of *how* you arrive at your numbers.

Finally, compare daily mineral intake with RDIs recommended by the federal government. Do adolescents have adequate mineral intake? Are there deficiencies in specific minerals? Are mineral supplements necessary? If there are deficiencies, what kinds of food could provide the missing minerals? Are there gender differences in mineral intake among adolescents? Is mineral intake more related to *amount* or *type* of food eaten? Naturally, if you do find a deficiency in someone's diet, advise the person to tell his or her parents. Help from a nutritionist or doctor may be called for. Do not attempt to correct the deficiency yourself!

It is entirely possible that the analysis of your data will reveal interesting new questions for you to ponder and pursue. Or you may wish to alter the original investigation in one of the following ways:

- Are adults any more likely to take in the necessary amounts of minerals than adolescents? Carry out a comparison study to find out, asking a group of adults and a group of adolescents both to keep food diaries for you to analyze.
- Pick any of the following to compare for the mineral analysis of their diets: males and females of the same age, those who eat at home regularly and those who do not, two ethnic groups which eat traditionally different diets, elderly people who have a meals program and elderly people who cook for themselves, health food enthusiasts and traditional eaters, vegetarians and meat eaters.
- Does any one food item contribute greatly to the mineral content of a diet? Does the mineral

content of the water or other liquids one consumes contribute greatly to dietary mineral intake? Which minerals, if any, are added to or removed from municipal water supplies and why? Does water softening significantly affect the mineral content of water? Is this beneficial to the diet?

ANTIBIOTICS AND FOOD SAFETY

There is widespread use of antibiotics in American agriculture. Antibiotics are substances, such as penicillin, which kill bacteria and other microorganisms. When antibiotics are used, they kill the strains of bacteria which cannot resist their effects, but sometimes a few resistant bacteria are not killed. Thus, the frequent use of antibiotics encourages the growth of antibiotic-resistant strains of bacteria. The concern is that when antibiotic-treated food is consumed, resistant bacteria in the food could pass on their resistance to other bacteria in the human body. If this happened, the antibiotics people use for disease might lose their effectiveness.

Are antibiotic-resistant bacteria present in or on the fruits and vegetables you buy? You can answer this question by rubbing produce items with sterile swabs and streaking the material onto agar plates containing antibiotic-impregnated paper disks as shown in Figure 3.

Streak the swab lightly but firmly over the agar plate making a zigzag pattern on the plate. To streak a plate properly lift the cover at an angle with one hand, while you streak the plate with the other, as shown in Figure 4. Do not remove the cover completely. Tape all the petri plates closed and incubate for 48 hours in a 37° C incubator or other consistently warm spot. A heating pad set on low substitutes nicely for an incubator.

After 2 days, if antibiotic-resistant bacteria are present, they will be able to grow right up to the disks. Bacteria will

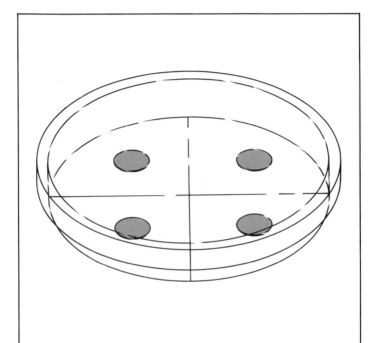

Figure 3. For the project testing for antibiotic-resistant bacteria in fruits and vegetables, the bottom of the petri plate is marked off into four quadrants. A different antibiotic-soaked disk is then placed in each.

not grow in the area immediately around each disk that contains an antibiotic to which they are not resistant.

Safety Note 1: Never open petri dishes which contain unknown bacteria. Always wash up with a disinfectant soap and plenty of warm water. All used petri plates should be autoclaved or cooked in a pressure cooker before being placed in marked plastic bags for proper disposal.

Safety Note 2: Do not experiment with animal products unless your science adviser approves. Some bacteria,

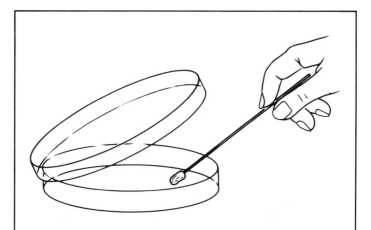

Figure 4. When streaking a petri plate with microorganisms, lifting the lid only partway helps prevent accidental contamination from the air.

such as salmonella, can be picked up from chicken and make you seriously ill or even send you to the hospital.

WHAT'S A HEALTHFUL DIET?

Because the snacks and meals we eat may contain hundreds of different chemicals, it is difficult to make specific and certain diet/disease connections. There are a few clear associations such as the one between calcium deficiency and osteoporosis, a deterioration of the bones. While other links are less obvious, it is generally acknowledged that heart disease, certain cancers, high blood pressure, and diabetes are linked to diet.

You can investigate aspects of a healthful diet by using the food diary and USDA data described earlier.

• Analyze the school lunch menu for a week to

determine whether it provides a significant portion of the RDIs. Check with the food service manager to find out whether RDIs are used in planning menus. If your results show that less than a third of several RDIs are being provided, what will you do with the information you have learned?

- How healthful are fast-food meals? Is the new lean hamburger being promoted by one fast-food outlet more healthful than a taco or chicken meal from another franchise? Fast-food restaurants will provide detailed nutritional analyses of their offerings on request. Conduct a literature search on diet links to heart disease and high blood pressure. Then make recommendations on the relative healthfulness of fast-food choices.

- Does the number of calories consumed on a regular basis affect longevity? Select an easily observed invertebrate species with a short life-span, such as the water flea *Daphnia*. Learn the recommended food types and amounts from a source such as the *Carolina Protozoa and Invertebrates Manual*. Design a controlled experiment in which you test the effects of calorie intake on longevity. Test low-calorie, high-calorie, and recommended-calorie diets. Look at several *Daphnia* regularly for observations of possible differences in function and behavior resulting from the different calorie intakes. Refer to Chapter 7 for more information on the care of *Daphnia* and for a photograph.

- Fiber has been promoted as possible aid in lowering risk of colon cancer. The water-insoluble forms, those containing cellulose or lignin, are thought to be particularly helpful. Design a test comparing the water absorbency of various types of dietary fiber.

EATING EXTREMES

Being as little as 10 to 20 percent overweight can increase a person's risk of certain cancers, high blood pressure, and heart disease. At the same time some of the behavior associated with excess dieting and underweight is equally life-threatening.

- Does the person who thinks he or she needs to lose weight always really need to? Design a study which compares adolescents' actual weight status with how they view themselves. Use the Metropolitan Insurance Company height and weight tables to determine actual weight status.
- Are adolescents more critical of their own appearance, including weight, than they are of the appearance of others? Use photographs of strangers who are overweight, normal weight, and underweight to investigate this question.
- Do reducing diets provide the RDIs in their menu plans? Do reducing powders? Compare several popular reducing diets to determine whether they meet these standards.
- Find out why so many people who lose weight on diets regain it.
- Investigate and compare the professional training and qualifications of individuals who provide diet services for clinics and weight loss centers.

READ MORE ABOUT IT

Ballas, Frank, Frank Fazio, Geno Zambotti, J. J. Costa, and Patricia Metz. *Physical Science with Environmental and Consumer Applications.* Dubuque, Iowa: Kendall Hunt, 1983.

Brody, Jane. *Jane Brody's Nutrition Book.* New York: Norton, 1981.

Burns, Marilyn. *Good for Me!* Boston: Little Brown, 1978.

Carolina Protozoa and Invertebrate Manual. Burlington, N.C.: Carolina Biological Supply Company, 1980.

Diet and Health Implications for Reducing Chronic Disease Risk. Washington, D.C.: National Academy Press, 1989.

Franz, Marion J. *Fast Food Facts.* Wayzata, Minn.: DCI, 1980.

Garrison, Robert H. Jr., *The Nutrition Desk Reference.* New Canaan, Conn.: Keats, 1985.

Mallick, M. J. *Anorexia,* Burlington, N.C.: Carolina Biological Supply Co., 1987.

Newton, David E. *Consumer Chemistry Projects for Young Scientists.* New York: Watts, 1991.

Nutrition and Your Health: Dietary Guidelines for Americans. Home and Garden Bulletin 228. Washington, D.C.: United States Department of Agriculture, 1990.

Nutritive Value of Foods. HG-72. Hyattsville, Md.: Human Nutrition Information Service, United States Department of Agriculture, 1985.

Perl, Lila. *Junk Food, Fast Food, Health Food.* Boston: Houghton Mifflin, 1980.

Raloff, J. "Low-Magnesium Diet May Clog Heart Arteries." *Science News,* April 7, 1990, p. 214.

Recommended Dietary Allowances, 10th ed. Washington, D.C.: National Research Council. National Academy of Science, 1989.

Saltman, Paul, Joel Gurin, and Ira Mothner. *The California Nutrition Book.* Boston: Little, Brown, 1987.

Scarpa, Ioannis, Helen Kiefer, and Rita Tatum. *Sourcebook on Food and Nutrition.* Chicago: Academic Press, 1982.

4

BRAIN AND MIND

When the lights dim in a movie theater, many possible experiences await you: fantasy, intellectual stimulation, fear, excitement. One organ, wholly or in part, controls your response and experiences. It is the site of creativity, sexual interest, anger and love, even the hunger and thirst that drive you to the concession stand for popcorn. This organ is your brain.

Your brain with its billions of neurons is the most complex organ of your body. Much is already known about the nature of nerve cell connections and the chemistry which control them. Yet, even if scientists were able to map every one of the trillions of neural connections, many people believe that we would not yet understand all of the human mind. For the mind, while governed by the brain, is not the same as the brain. The mind both depends on the brain and adds an interpretive dimension to the brain's functioning. It evaluates, learns, forms memories, interprets outside events, analyzes pain, and so on.

This chapter suggests several ways to investigate a small

part of the complex organ, the human brain, and a complex concept, the human mind. You are a member of the only species able to study its own brain. What are you waiting for?

THE BRAIN AND LEARNING

What is the best way to learn? If the learning process were easier to define, it might be possible to help more people learn in a more effective manner. Acquiring information and skills, processing, storing, making connections, retrieving information, solving problems—these are among the many facets of learning. Scientists don't really understand all the hows and whats of learning, so many areas are open for investigation.

You can begin your study of learning with one of the short projects described here. Start by trying the investigation on yourself and a friend. Later you may want to expand your investigation to a larger group.

Guidelines for Working with Human Subjects: See Chapter 1.

Under what conditions does a person learn the most? Select a learning task first, for example, learning five new words and their meanings. Using a dictionary, select several groups of five words unfamiliar to most people. To limit variables, choose all words with the same number of syllables, and which all have short definitions, say, fewer than ten words long.

- Give your friend the list of five words and their meanings to read over. After 10 minutes, ask her or him to write the words and their definitions. Then give her a new list and instruct her to copy the words and their meanings while trying to learn them. Again test her 10 minutes later. Finally give her a third list and ask her to repeat the words and their meanings over and over. Test her after 10 minutes. Depending on the results and the person you choose, you may have to lengthen the list or the amount of study time. Does the method of study—reading, writing, or saying—influence the number of words learned?

- Which method produces the fastest learning? Use the lists, but this time keep testing every 10 minutes until your friend has learned all of the words. This will allow you to determine which method is fastest.
- Does combining two methods, for instance, writing the definitions and saying them at the same time, speed the learning process?
- Does moving the body, for example, by walking around the room, affect learning by reading or saying?
- Speak five new words and their meanings to a friend. Repeat the words to her and ask her for the meanings. Next speak five additional new words to your friend while she closes her eyes. Again ask her for the meanings. Does closing the eyes affect learning? Would closing the eyes while responding help your friend better recall the answers?
- If you find that differences in learning emerge from any of the projects, try to find out whether these differences are repeatable with the same person by doing the same type of test with that person on different days. Also, do other people show the same "best" ways to learn new words?
- Do "best" ways to learn change if the type of learning task changes? Try similar methods on a list of words in a foreign language. Test a physical skill with several steps by asking some subjects to read about it, some to talk about it, and some to hear about it at the same time they perform the sequential task.
- It is possible to learn new words and to forget them later. Test your subjects right after they have

learned new words, several hours later, and several days later. Does the method of learning—speaking, writing, saying—affect how long information is retained?

MEMORY AND LEARNING

Can drinking a high-glucose liquid enhance memory? Some tests with elderly volunteers seem to point in that direction. Is this also true for younger adults or for adolescents?

Guidelines for Working with Human Subjects: Make sure that your subjects have no health restrictions on sugar intake.

You can start by testing the idea with a friend. Give your friend a drink sweetened with glucose. (Corn syrup is an inexpensive, concentrated source of glucose.) Forty-five to 90 minutes later, give your friend a memory test. The test can be a list of twelve two-digit numbers of twelve unrelated words. Speak the list to your friend. Ask him to repeat the list and give him several chances to "pass" the test by repeating the list after each of several tries. Score him on the number of attempts required to repeat the list of words or numbers correctly.

Repeat the test on another day with the same friend. The second time, give him a drink, such as mineral water or an artificially sweetened beverage, which does not contain glucose. Give the same size drink used before. Again, wait 45 to 90 minutes to administer the test, using a different list of the same length and difficulty as the first. Is there a difference in the number of tries to learn the list with and without glucose?

- Will you get different results if different sugars, such as sucrose (table sugar), lactose (milk sugar), or fructose (fruit sugar), are used?
- Does the amount of time that has passed after a meal affect results?

- On the basis of results for one person, can you predict what might happen in a larger study? Carry out the test as described earlier with ten to twenty volunteers.

Here are some additional ideas you may wish to investigate whose focus is memory and/or learning.

- Are facts learned just before the brain rests retained better than those learned sandwiched between other mental activities? Do some tests of learning just before sleeping or meditating and compare them with learning attempted between two problem-solving sessions or two unrelated tests.
- Does background music affect memorization? Compare either speed or accuracy of memorizing with and without background music.
- Do people remember more from a supportive teacher?
- Does smell affect memory? Conduct memorization tests as described, but do so in a room with a pleasant and noticeable odor, such as that from peppermint or baking cookies. Provide the same pleasant odor to check half your group several days later for the effect of odor on recall of information. Do not provide the odor for the other half of the group and compare recall scores.

STRESS

Stress is often painted as a villain—the cause of health problems, impaired judgment, and decreased enjoyment of life. Major stresses such as divorce, death of a loved one, moving, and loss of a job do lead to a higher than average rate of medical problems. However, the role of stress in daily life is not entirely negative.

Stress occurs when an external factor causes external changes in the body, upsetting the normal balance. The body then responds in a variety of physical and chemical ways. These responses can help you focus attention on a specific task and direct blood sugar to your brain and muscles. Other stressful situations may have a negative impact. Many students consider time limits in testing stressful. Do these time limits stress in a way which helps or hinders performance?

To investigate this question, generate several versions of a sheet containing the numbers 100 to 199 randomized on a piece of paper. Table 5 shows three versions of such a sheet. No pattern should be apparent as one looks at the paper. Test two groups of randomized volunteers of approximately the same age using these sheets of numbers. Give each group the following directions:

Inform them that you are giving them a test of eye-hand coordination. When they receive their sheets of paper, they are to circle the numbers 100 to 199 in numerical order working as quickly as possible. Tell *only one group* that there is a 2-minute time limit. Conduct the two tests under identical conditions. Tell *both* groups to stop after 2 minutes. Compare scores statistically using means, standard deviations, and a test of significance, as explained in Chapter 1.

Do you find a significant difference when people are told ahead of time that there will be a time limit?

- Does the perceived shortness of time make a difference to outcome? Test this idea by announcing a 30-second time limit instead of a 2-minute limit.
- Does gender play a role in stress effects on task completion?
- If the task was to learn a list of unrelated words, would an announced time limit affect learning ability?

TABLE 5. RANDOM NUMBERS FOR THE STRESS PROJECT

171	119	183	114	122	187	148	141	101	194
117	151	145	135	165	184	134	193	102	162
188	190	160	166	170	185	180	118	105	100
110	195	186	138	115	130	112	177	169	109
173	150	155	196	197	142	168	147	137	175
152	106	179	108	178	131	157	143	191	123
172	182	159	167	107	156	199	129	153	161
176	154	124	189	144	127	136	104	126	174
198	113	139	146	103	164	149	133	120	132
121	192	111	116	158	163	125	181	128	140
109	135	118	161	144	174	196	165	143	175
123	120	199	168	188	110	195	106	114	124
130	104	129	139	179	119	145	198	162	147
156	157	151	117	140	170	125	186	154	108
163	167	138	103	132	122	184	169	131	192
181	111	128	173	191	141	146	121	127	115
134	155	193	171	148	158	176	137	100	152
159	189	102	160	107	116	149	177	101	153
178	185	182	142	133	197	183	180	136	126
166	190	113	172	105	150	187	194	112	164
122	125	150	168	157	107	175	124	181	115
129	187	188	161	110	183	123	170	101	189
196	103	111	191	143	184	102	165	199	118
113	154	104	174	159	194	182	137	138	180
185	169	135	127	151	179	100	166	160	197
147	167	114	145	146	142	132	149	171	112
134	164	130	173	148	176	195	121	158	152
139	178	126	106	198	156	162	144	116	108
128	153	193	190	186	172	192	109	133	105
163	177	120	141	117	131	119	140	155	136

OTHER FORMS OF STRESS

Additional forms of stress have possible effects on task performance that can be assessed by the same general experimental procedure outlined above:

- What is the effect, if any, of irritating behavior on task performance? Expose one of your test groups to popping gum, drumming fingers, pencil tapping, or some other annoying type of behavior and compare performances. It is especially important that the groups *not* know what you are testing in this type of situation.
- Does background noise affect task performance?
- Does music affect task performance in a way an individual subject can predict? Ask subjects to select music they think helps them study and use that music in this test situation.
- Does the promise of a reward alter task performance?
- Does announcing the expectation that students will do well have a different effect on task completion than announcing that they are expected to do poorly?

For each of these situations you may modify the nature of the test given to differentiate between tests of manual dexterity and more thoughtful tasks, such as adding lists of numbers, picking out matching objects, and repeating lists of words. Does the same variable affect both kinds of tests similarly?

HEMISPHERE DOMINANCE

The brain has two halves, called hemispheres. These two halves are connected by a thick bundle of nerve fibers

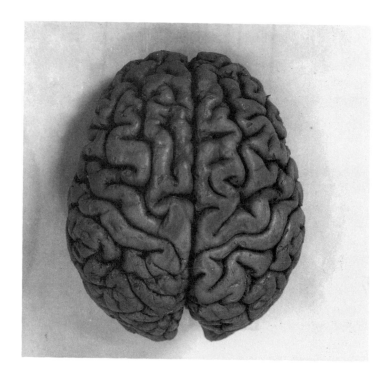

Photo 6. The human brain is divided into two hemispheres, each specializing in different functions.

called the corpus callosum. Photo 6 shows you just how distinct the two halves of the human brain really are. About 98 percent of the adult population is left-hemisphere dominant for reading, speaking, and other verbal activities. Is this true of children also, or is hemisphere dominance acquired during maturation?

Test this question by creating a situation in which children under the age of 8 years view a set of composite photographs. Construct the photographs in the following manner: Get ten to twenty subjects, some male and some female. Using black-and-white film, photograph each subject's face with a pleasant expression on it. Then, without changing the subject's position relative to you as photographer, direct the subject to appear angry. Take a second photograph. Have the photos developed. Cut the two photos of each person down the center of the face and recombine them so that each has a happy and an angry half. This is probably the most difficult part of the project. Careful technique, so that each composite photo looks like a single face, will enhance the success of this project.

Show each child the series of ten to twenty composite photographs straight on and ask him or her to rate the faces as either happy or angry. Because information seen on the left is interpreted by the right side of the brain and vice versa (Figure 5), if the left hemisphere is dominant, the child will verbalize what he or she has "seen" on the right side of the photo. If your results show that the overwhelming majority of children are left-hemisphere-dominant, as adults are, offer a hypothesis on the origin of this dominance. Regardless of outcome, you may wish to investigate related questions such as the following ones:

- Do most adolescents show left hemisphere dominance?
- Is handedness correlated with hemisphere dominance?
- Some studies show less dominance and more "crosstalk"—more neural activity between the hemispheres by way of the corpus callosum—in female brains than in male brains. Can you find such gender differences? If brain dominance does differ in males and females, at what age does this difference arise?
- Can most people learn a list of numbers better if it is presented to the subject to the right of center of the visual field than if it is presented to the left of center of the visual field? Try presenting lists of ten numbers separately to the right

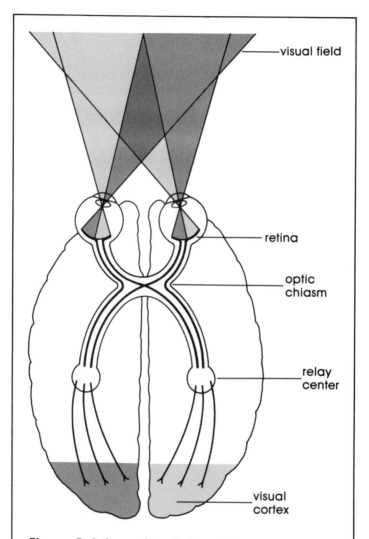

Figure 5. Information in the left visual field, although received by both eyes, is transmitted only to the right side of the brain. Similarly, information from the right visual field reaches only the left side of the brain. Source: H. Curtis Biology, Worth, 4th ed.

and to the left on pieces of paper. Subjects must look at the page straight on, should not be allowed to move their heads, and must repeat the lists after seeing them for 10 seconds each. When designing this experiment, remember to take into account the effect of training. One often gets better at a task as it is done more times!

- Is one eye more important than the other in remembering a list of numbers presented visually? Design a variation of the test to look at this question.
- Construct your own tests of hemisphere dominance using artwork or simultaneous slide projections on two screens if equipment is available.

READ MORE ABOUT IT

Gazzaniga, Michael S. *Mind Matters.* Boston: Houghton Mifflin, 1988.

Haines, Gail K. *Brain Power: Understanding Human Intelligence.* New York: Franklin Watts, 1979.

Lambert, Mark. *The Brain and Nervous System.* Englewood Cliffs, N.J.: Silver Burdett Press, 1988.

LeDoux, Joseph E. and W. E. Hirst. *Mind and Brain.* New York: Cambridge University Press, 1986.

Mishkin, Mortimer, and Tom Appenzeller. "The Anatomy of Memory." *Scientific American,* June 1987, pp. 80–89.

Restak, Richard M. *The Mind.* New York: Bantam, 1988.

"Sweet Remembrances." *Science News.* September 22, 1990, p. 189.

Wurtman, Richard J. "Nutrients That Modify Brain Function." *Scientific American,* April 1982. pp. 50–59.

5

THE NATURE OF DISEASE

Sniffles, a stuffy nose, and assorted aches and pains are among the symptoms that often signal the arrival of a cold. Although a cold is considered a minor disease, like all disease it diminishes the body's ability to function. We know that viruses cause colds. We know that time and rest are the treatments. But we are still a long way from knowing how to prevent colds and how to recover from them quickly.

What we do know about a disease is focused in three main areas: what causes them, how you get them, and how they can be prevented or treated. We have learned that disease in humans can be caused by diet deficiencies, environmental factors such as poor sanitation and pollution, aging, life-style choices, heredity, and other organisms. Many of these work together to produce their effects. For example, cigarette smoking puts a person at risk for lung cancer, but hereditary factors probably also play a role.

The kinds of disease best understood and most successfully treated are those caused by other organisms, including bacteria, fungi, and protists. Organisms that cause disease are known as pathogens. Unlike the cold virus, these agents of disease can often be treated successfully with antiobiotic drugs, which kill the organisms.

Safety Note: In projects involving microorganisms, sterilize all used culture plates in an autoclave or pressure cooker and dispose of them properly in marked biohazard bags.

HOW A DISEASE IS SPREAD

How do we catch cold? Is it by standing in a draft getting chilled, being sneezed on, being worn out, not getting enough sleep? And how can colds be prevented? By staying warm and dry, drinking lots of orange juice, washing your hands a lot? Being conscious of touching people who have colds, or using their tools, telephones, or computers and not touching your mouth, nose, or eyes?

You can test the handwashing hypothesis, but before you do, be sure to familiarize yourself with basic bacteriological techniques, including proper culturing, handling, and disposal techniques.

Safety Note: Work under supervision, using only the microorganisms suggested or substitutes which do not cause human disease. Do *not* attempt these activities with unknown microorganisms.

Obtain a culture of *Vibrio fischeri,* an organism harmless to humans, from a biological supply house. Either prepare or purchase at least a dozen photobacterium agar culture plates. When you culture the bacteria in this agar, they will emit light, a property called luminescence. You then can easily detect the *fischeri* colonies.

Assemble three or four volunteers to whom you have explained the nature of the experiment. Ask them to wash their hands with a disinfectant, such as diluted Lysol, and plenty of warm water and let their hands air dry.

Assign consecutive numbers to your volunteers and

have them stand in line according to number. Instruct the first person to place his or her fingers on the *Vibrio* culture. This first person should then shake hands with the second person. The second person shakes hand with the third person, and the third with the fourth.

Each person then touches his or her fingers to the surface of a sterile plate of photobacterium agar and immediately closes the cover. Each volunteer should again disinfect his or her hands.

Mark each plate on the bottom with the number of the individual who touched it, tape it shut, and store it upside down at 25° C.

Look for evidence of *Vibrio* growth after 1, 2, and several days. Were the bacteria transmitted? On which plate or plates can the greatest number of colonies be found? Are there bacteria on the highest-numbered plate? If so, how could you improve your original test?

Next, test the effectiveness of the suggested hygienic technique by asking the second or third person in the chain to wash his or her hands in soap and warm water after receiving the "infected" handshake but before shaking the next person's hands. How well does simple hand washing interrupt bacterial transmission? Does the length of time spent washing hands affect the outcome? Does the temperature of the water make a difference? Which disinfectant soaps or liquids are effective, if any?

Here are some other questions to investigate:

• Does the species of bacterium involved affect efficiency of hand-to-hand transmission? Other species easily detectable on agar are *Micrococcus luteus,* which produces a yellow pigment on ordinary bacterial agar; *Micrococcus roseus* and *Rhodococcus rhodochrous,* which produce a rosy pigment; and *Serratia marcescens,* which produces a red pigment. All can be grown on bacterial agar at 25° C.

VIRUS TRANSMISSION

Are viruses effectively transmitted hand-to-hand? Obtain a culture of *Escherichia coli* bacteria. Prepare many plate cultures by spreading the *E. coli* from a broth culture onto agar plates.

Use sterile equipment throughout this experiment. Tape, invert, and grow the cultures at 37° C for 24 hours.

Obtain a culture of coliphage, a virus which infects and kills *E. coli.* Arrange your volunteers as before. This time the first person places her fingers in the coliphage culture and the handshakes proceed. After the handshakes are finished each person then places her fingers in a growing culture of *E. coli.*

Safety Note: Volunteers should wash their hands thoroughly with a disinfectant soap immediately after touching the bacterial plates. Do not permit people with recent cuts or abrasions to participate.

Keep the *E. coli* cultures inverted at 37° C for several days. If any clear areas show up on the cultures, the virus has been transmitted to the culture. Is the virus transmitted hand-to-hand? On what plate or plates are the greatest number of clear areas found? Is any kind of pattern apparent? What would the presence of clear areas in the highest-numbered plate indicate? Should you do more extensive testing?

Here are some additional ideas to test:

- Does washing with ordinary soap and water stop viral transmission hand-to-hand?
- Will bacteria or viruses stay alive for 2 minutes, 5 minutes, or 10 minutes between handshakes? Devise an experiment along the same lines as the last two projects to test the "survivability" of bacteria and viruses on hands between handshakes. Set a reasonable interval between handshakes because volunteers will not be able to touch anything during this time.

- How long will a coliphage on a damp tissue remain infective?
- Are viruses or bacteria transmitted from other surfaces, such as countertops? Swab a clean surface with a small amount of coliphage culture. Wipe the surface again with a sterile swab 1 hour or more later and transfer to a growing *E. coli* culture. Look for clear areas in the culture dish after several days.

Safety Note: Clean all areas used in bacterial or viral experimentation thoroughly with a strong disinfectant and warm water. Use only microorganisms not harmful to humans.

- Many investigations similar to those described can be carried out using fungi (such as yeast) instead of bacteria or viruses. Several fungi, including yeast of some types, cause human disease.
- Which major group of organisms—bacteria, viruses, or fungi—seem most easily transmitted by hand-to-hand or surface contact?

DISEASE CARRIERS

Viruses, bacteria, fungi, and pathogens may be transmitted through the air; by skin contact; and also through soil, water, and food. In addition they may also be borne by vectors, organisms that carry the disease but are not themselves affected by it. One vector is the flea that can carry bubonic plague from rodents and other small mammals to humans. The flea is an insect which can transfer bacteria or other microorganisms by first biting an infected person and later biting an uninfected person.

In the fourteenth century, plague wiped out an estimated 25 to 65 percent of the population of Europe. The plague, caused by the bacterium *Yersinia pestis,* continues to be a public health problem in certain areas of the United States.

Some of the other major vectored diseases in the United States are Chagas' disease, Lyme disease, Rocky Mountain spotted fever, tularemia, babesiosis, toxoplasmosis, and rabies. The vectors are mammals or insects and ticks whose hosts are mammals. Libraries, public health agencies, and your local veterinary pathologist (associated with a university, county, or state agriculture department) may be able to assist you with the following kinds of questions or investigations related to vectored diseases:

Safety Note: Work under supervision and do not experiment with or risk infection from vectors that cause diseases like those discussed above.

- How often are family pets, such as cats and dogs, a factor in the transmission of diseases to humans? Is this a significant route of disease transmission? What precautions if any should pet owners take? Make an analysis of local, regional, or state statistics to look into these questions.
- Are any vectored diseases significant enough in your part of the country that public health measures are undertaken for prevention? Which methods of prevention are considered most effective and why? Do different areas of the country undertake different prevention programs? If so, has their effectiveness been compared or analyzed?
- Construct a country, state, or U.S. map of reported cases for one or more vectored diseases. Are the rates of vectored human diseases in your state or the entire United States changing? Is there a pattern? Can you identify reasons for the changes?
- What infective agents are carried by the ticks or mosquitoes in your area? What percentage of

the tick or mosquito population carries these pathogens?

CONTROLLING PATHOGENS

When prevention of disease fails, treatment is often necessary. Antibiotics are an effective treatment for the kinds of disease caused by bacteria, protozoa, and other pathogens. Mostly simply, antibiotics are substances that kill living things.

The first antibiotic used in the wide-scale treatment of bacterial disease was penicillin, a substance naturally produced by a mold. The search continues for new antibiotics today. Although many of today's antibiotics are synthetic, scientists still search for antibiotics made by living things. Wherever molds, a type of fungus, grow, there is the potential to find a new antibiotic.

Considering their often damp conditions, do soils contain molds? If so, do those molds have antibiotic capabilities? To answer these questions, you can undertake a series of steps which isolate molds from soils and then test them for their antibiotic properties.

Collect different kinds of soil samples: sandy, loamy, clay, with and without humus, etc. Collect soil in areas free of trash or known pollution. *Use gloves in collecting and handling all samples.* Make a suspension of each in distilled or spring water. Filter each with cheesecloth. You will need to separate mold spores from other forms of life in this filtrate. You can use the following method:

Using a sterile technique prepare a starch-yeast agar medium. Starch-yeast agar can be purchased, but if you wish to make it yourself from components, consult the *Difco Manual* (see the list of books at the end of this chapter) for directions.

Add two antibiotics to the agar. The concentration should be 1 milligram (mg) of each antibiotic for each 1 milliliter (ml) of agar. Use a combination, such as penicillin and tetracycline, which is effective against a wide variety

Photo 7. Four different types of antibiotics were tested for their ability to inhibit growth of the bacteria *Bacillus cereus.* The antibiotics were (clockwise from top): erythromycin, neomycin, tetracycline, and penicillin. To which antibiotic is the bacteria resistant?

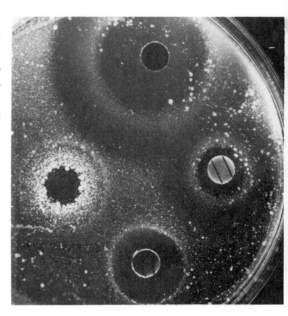

Photo 8. "Testing ground": pinches of soil, diluted in sterile water, were incubated in petri dishes containing nutrient agar. After several days, colonies of different organisms found in the soil appear on the agar. Scientists test these microorganisms for antibiotic properties. (Photo courtesy of Pfizer, Inc.)

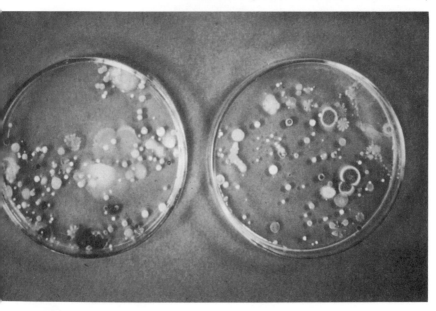

of bacteria. These antibiotics can be purchased through a biological supply house or a feed and grain store. Their purpose is to inhibit the growth of bacteria while you are trying to isolate molds from your soil extracts.

Pour the agar into sterile petri plates and cover. When the agar has set, streak several plates with extract from one soil source. Dip a cooled, sterile inoculating loop into the extract and streak patterns similar to those shown in Figure 6. Drag the loop lightly across the surface of the agar in a zigzag manner, rotate the dish 90 degrees, and repeat the zigzag. Then rotate it another 90 degrees and streak again. At the end, so little material should be left on the inoculating loop that single organisms may be found in the last few streaks.

Repeat the streaking procedure on several plates for each soil extract, with plates carefully marked to identify which soil source extract each contains. Invert the plates and incubate at 37° C for 12 to 24 hours. If molds are present in the soil samples and the single-colony isolation technique works as intended, you will find single, rounded, and fuzzy colonies of mold on some of your plates.

At this point, use another sterile inoculating loop to lift individual mold colonies off the plates and transfer them to individual sterile tubes of liquid medium (same yeast-starch base without antibiotics). Grow these tubes at 37° C for a week.

Meanwhile, prepare one group of bacterial agar plates streaked with *Bacillus subtilis,* a gram-positive bacterium, and another group of plates freshly streaked with *E. coli,* a gram-negative bacterium.

Using a hole punch, make tiny disks from highly absorbent purified grade filter paper. Soak the disks in your liquid mold cultures.

Drop two or three culture-soaked disks onto each bacterial plate and add a plain unsoaked paper disk to each plate as a control. Try to place the disks in different regions of the plate. Press the disks gently into place and

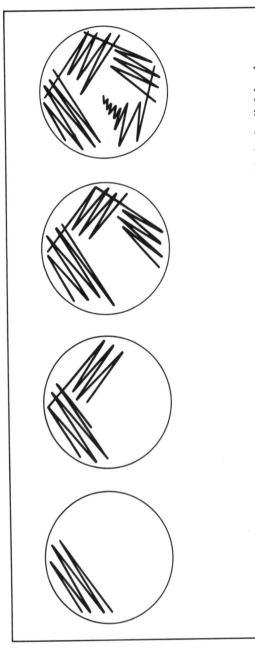

Figure 6. This series of drawings shows the technique used to obtain individual colonies of fungi for the project on testing fungi for antibacterial properties. Each time the dish is rotated, the inoculating loop is dragged across only a small part of the previously streaked area. Few organisms should be present in the last region streaked.

mark the bottom of the plate with information about the source of the material on each disk.

Incubate the inverted plates at 30° C for 24 to 48 hours. If any of your mold cultures has antibiotic properties, bacteria will *not* grow in the area immediately around the disk soaked in that culture. In fact, clear circular areas will appear around disks that have antibacterial properties. You can answer a number of questions with the results from this experiment.

Are there antibacterial molds in the soils you tested?

If so, what type of soil—wet or arid, high or low in humus content, cultivated or uncultivated, acidic or basic—best fosters antibacterial mold growth?

If antibacterial molds are in the soil, are they more likely to be effective against gram-negative or gram-positive bacteria?

Can any antibacterial molds be found which are effective against both gram-positive and gram-negative bacteria?

Some related questions you can answer with modifications of this experimental procedure are the following:

- Are leaf piles, compost, and grass clippings good sources of antibacterial molds?
- Does pH affect isolation, growth, or effectiveness of antibacterial molds?
- Does a combination of antibacterial molds have an additive effect in inhibiting growth of bacterial cultures?
- It has been suggested that for evolutionary reasons tree bark and seed coats might contain antimicrobial properties, effective against bacteria, or fungi. Design an experiment to test this idea.
- You may have another idea about a natural source of antimicrobial agents. Form a hypothesis and test it with the techniques described.

OTHER CAUSES OF DISEASE

Many of the links between hereditary factors and other agents of disease appear to be complex. For example, we are currently unsure of the connections among stress, diet, heredity, and exercise level in the development of heart disease. What seems to be clear is that each of these components is important to overall health. That is why each of them is the subject of another chapter in this book. Do look further for even more ideas!

READ MORE ABOUT IT

Burns, Gwendolyn. *Microbiology for the Health Sciences.* New York, Lippincott, 1988.

Cobb, Vicki. *Lots of Rot.* New York: Lippincott, 1981.

Cooper, J. I. and F. O. MacCallum. *Viruses and the Environment.* Cambridge: Cambridge University Press, 1984.

Difco Manual of Dehydrated Culture Media and Reagents for Microbiological and Clinical Procedures. Detroit: Difco, 1953.

Dubos, Réné. *The Unseen World.* New York: Rockefeller Institute Press, 1962.

Flint, S. Jane. *Viruses.* Burlington, N.C.: Carolina Biological Supply Company, 1988.

Gest, Howard. *The World of Microbes.* Madison: Science Tech. 1987.

Gutnik, Martin. *Immunology: From Pasteur to the Search for an AIDS Vaccine.* New York: Franklin Watts, 1989.

Siegmund, Otto, ed. *Merck Veterinary Manual,* 6th ed. Rahway, N.J.: Merck, 1985.

Teasdale, Jim. *Microbes.* Morristown, N.J.: Silver Burdett, 1984.

6

ENVIRONMENTAL HEALTH ISSUES

Two thousand years ago the Romans recognized the health dangers posed by exposure to lead. The lead pipes that carried their water caused widespread poisoning and behavior that they described as "madness." In the Middle Ages the English debated restrictions on peat, coal, and wood burning because of dense air pollution and suspected hazards. Today we recognize that no chemical substance is completely safe. All are toxic at some level, even water in very large amounts. The differences among substances are a matter of the extent of exposure and of the amounts which are hazardous.

Environmental threats to health are real but often hard to pin down. Because a person is exposed to literally thousands of substances, the job becomes one of separating out the effects of these substances. In addition, one of the greatest needs is to find better ways to collect information about exposures and effects.

What are some of the hazards of concern in our environment? What effects do they have on living things? Does exposure to these hazards occur only in chemical plants, mining operations, and dumps? Are there toxics close to home? Getting involved in one of the projects described in this chapter could help you with these questions.

REAL VS. PERCEIVED RISKS

People frequently overestimate or underestimate the actual risk of a situation, activity, or exposure. Even when a risk is known, behavior may not be modified to match the severity of the risk. For example, the same person who won't eat sugared cereal may routinely ride in a car without wearing a seat belt.

Education may seem to be a logical first step in altering understanding of relative risks. Relative risk compares risky situations and determines which is more hazardous. Do people's abilities to gauge relative risk correctly vary with their level of education? You can answer this question with a survey.

Research twenty to thirty causes of death in the United States. Include the top ten, as well as some less significant causes of death. Make a list and ask your survey group to rank the causes of death from highest (causing the greatest number of deaths per year) to lowest. Be sure to survey a group of respondents at different education levels: non—high school graduate, high school graduate, some college, college graduate, etc. Try to survey as many people as possible at each level. This helps improve the chances that you will obtain meaningful data. Follow guidelines outlined in Chapter 1 for working with human subjects, but do not tell subjects the purpose or content or the survey before it is administered to them. After you collect the data, there are several ways to analyze it. One suggestion is given here.

Devise a scoring system which sums the differences between respondent rankings and the actual rankings. See Table 6.

TABLE 6.			
Causes of death	Actual ranking	Respondent ranking	Difference
A	3	2	1
B	2	3	1
C	4	4	0
D	1	1	0
		Total difference = 2	

Column 1 lists four causes of death. In your table there would be more than these four, and you can use the actual descriptions rather than letters. The second column contains the real ranking of causes of death. In this example, cause D is the number 1 cause of death. The respondents should not see the information in this column.

Column 3 gives the respondent's ranking. It is compared to the number in column 2, and the difference is placed in column 4. This respondent has a total score of 2, found by adding all the numbers in column 4. With this method of scoring, a small total means a more accurate assessment of risk and a large total reflects a less accurate assessment.

Compare mean scores for the education groups included in your survey. If there are differences, are they explainable by chance alone, or is three a high probability that education does play a role in this assessment of risk? Does more schooling correlate with better risk assessment? Does a college education guarantee a good understanding of risk? Apply the appropriate statistical test of significance to find the answers to these questions.

Here are some related questions about risks which you might wish to investigate:

• Does a strong technical or mathematical background help a person recognize and assess relative risk more accurately?

- Do causes of death reported most often in the newspaper tend to be overestimated in surveys of risk? For example, murders, airplane crashes, and automobile accidents are prominently reported by newspapers. Keep track of the number of deaths and their causes as reported in your newspaper for several weeks. Then conduct a survey of regular readers of the newspaper similar to the one described earlier. Modify it to include the most frequently reported causes of death in the newspaper as well as any of the top ten not mentioned. You can modify this project to include radio and TV, too.

- Do people avoid risky behavior when they know that the risk is significant? Provide information about a known risk to health, such as not wearing seat belts, to a group of people who can be observed for this behavior. Observe their seat belt habits before providing the information, set up a time to provide the information to the group, and then observe their seat belt habits after the education session. You may wish to experiment with several methods of education such as a lecture about statistics, an audiovisual program comparing outcomes with and without seat belts, first-person testimonials about the benefits of seat belts from crash survivors, etc. If your school's driver education program includes education on seat belt use, you might work with the instructor to monitor the effectiveness of its seat belt education component in affecting behavior. Because people will not always be truthful about their habits, after-school parking lot evaluations can be an effective way to learn about seat belt habits.

- Is the perceived risk of dying in an automobile

accident the same or different among teenage drivers, teenage nondrivers, and adult drivers?

WATER POLLUTION

One of the greatest environmental health concerns of recent years has been cleanliness of water. A variety of potentially harmful chemicals get into our water through industrial operations, farming and waste disposal, and home lawn care. When a chemical is being evaluated for possible harmful effects, one of the first tests undertaken is called a screening-level toxicity test.

You can investigate toxicity as it applies to lawn fertilizers and the health of invertebrates found in the soil and bodies of water near your home and other homes in your area.

Collect many specimens of one or more invertebrate species in the soil or water near your home. Do not collect from known polluted areas. Among the species you might use are various earthworms, ants, copepods, leeches, rotifers, and isopods.

You will also have to be able to keep the invertebrates alive long enough to do toxicity tests on them. This means you will need to know how to properly maintain the species you elect to study. One useful reference is listed at the end of the chapter. If the opportunities to collect small invertebrates in your neighborhood are restricted, you can order *Daphnia* or rotifers from a biological supply company or obtain earthworms or crickets at a bait shop.

A substance is considered acutely toxic to an animal species at a given concentration if that concentration kills 50 percent or more of an exposed group within 4 to 14 days of administration.

Once you have your specimens, apply a lawn fertilizer which does not contain pesticides. Test various doses of the fertilizer mixed in water with your invertebrate species of choice by adding these different doses to the soil or water in which the organisms are kept. Be sure to use fresh specimens of invertebrates each time you test a new toxicity level or dose. Invertebrates already exposed to a chemical may react differently from those not tested before. Keep control groups of each species not exposed to any fertilizer at all.

Does the fertilizer produce lethal effects in any of the species you tested? If so, what amount of fertilizer kills 50 percent of your specimens? Are there different toxicity levels for different organisms? What can you conclude about the use of the tested fertilizer and the invertebrates in the surrounding soil or water? What areas of your community use the greatest amount of fertilizer? Is your community's drinking water possibly affected by lawn fertilizers?

You can also test the effects of the following substances:

- Detergents—phosphate and nonphosphate
- Aluminum, which increases in the soil with rising acidity of rain
- Used motor oil, which is frequently disposed of improperly
- Outfall from three-stage wastewater treatment facilities
- Copper, a toxic material which enters water from mining and smelting operations

Safety Note: Work with the supervision of a qualified science teacher or other science professional. Wear approved safety goggles, protective gloves, and a lab apron. Use dilute solutions only, and treat all substances as if they are toxic.

- Seed germination tests can provide relatively quick information about toxicity of substances in the environment. These tests look for the effects of a substance on the ability of seeds to germinate and develop properly into young plants.

While the information gained from such experimentation does not necessarily apply directly to humans, the health of the environment is, in the end, a health concern of humans, too. When you make tests on seed germination, be sure to keep control data.

Keep in mind that, while finding that a substance's negative effect on health of an invertebrate, seed germination, or another life process may be significant, the *amount* of the substance required to effect the change is often just as important a question in risk assessment.

TOXICS AND THE UNBORN

The most serious incidence of mercury poisoning in humans occurred in Japan in the 1950s. Forty-three residents of Minamata Bay died as a result of eating mercury-contaminated fish. Pregnant women who ate the contaminated fish gave birth to babies with defects. Mercury and other agents that cause developmental abnormalities are called teratogens. Radiation and alcohol are other well-known human teratogens. While most agents known to be teratogens also harm adults, the risk may be magnified in the rapidly growing embryo or fetus. The gestation period is a time when cells are dividing and organizing themselves many times more quickly than they do in adults.

Do any of the metals with properties similar to mercury's, such as chromium, found in industrial wastes, or silver, used to disinfect swimming pools, have effects on rapidly growing organisms? One way to test this question is with an invertebrate organism capable of regeneration. Regeneration is the ability to regrow lost or cut parts of the body. It is characterized by the same rapid rate of cell division found in embryonic development. A good candidate for such work is the flatworm, *Dugesia tigrina,* shown

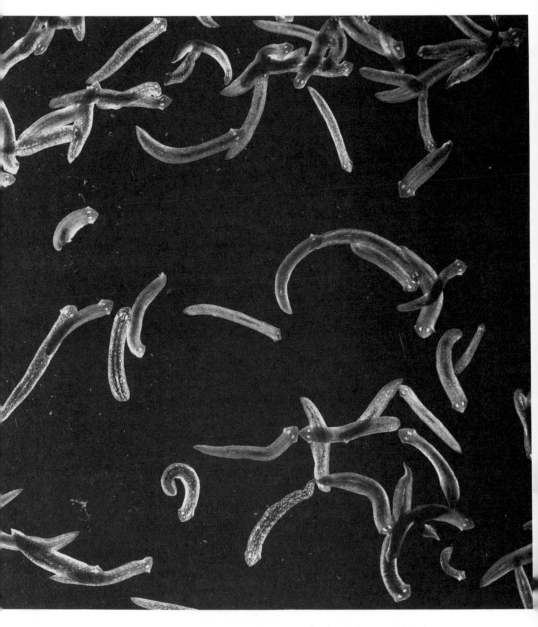

Photo 9. A planarian is a useful subject
for experiments on the effects of
chemical pollution on organisms.

in Photo 9. These organisms—called planaria—are easily maintained if given clean spring water or pond water daily and fed small amounts of hard-boiled egg yolk.

As with any organism kept for experimental purposes, familiarize yourself with the planarian's care, feeding, and reproductive cycle *before* beginning the experiment. For instance, *Dugesia* respond poorly to handling during their reproductive cycles. Consult a manual containing information on care of this invertebrate; one is listed at the end of this chapter.

First maintain cultures of planaria for several weeks. Use a clean single-edge razor blade or scalpel (careful!) to cut a large number of planaria into two or more large pieces. Refer to Figure 7 for some suggested cuts. The planarian will extend itself, making the cutting easier, if placed on a cold surface. Keep records of the numbers and kinds of cuts as you place the pieces into several culture dishes. The culture dishes should contain spring or pond water to which have been added small concentrations of silver or chromium ions. *Use only very dilute solutions of chromium or silver salts as selected by a qualified science teacher or scientist.* As a guide, the U.S. Environmental Protection Agency allows 0.05 milligram/liter (mg/L) of silver or chromium ion in drinking water, so this amount is considered safe for humans. Will this concentration interfere with planaria regeneration? Will a larger amount have an effect? Be sure to maintain one or more control cultures for comparison.

Caution: work under the close supervision of a qualified science teacher or scientist (for example, a chemistry teacher). Do not do this project unless you can get adequate supervision and your adviser or parents or legal guardian approves. Wear safety goggles, protective gloves, and a lab apron. Use dilute solutions or test chemicals. Be aware that *some silver compounds can cause severe burns and are mutagenic—that is, they can cause mutations in organisms—and that many chromium compounds are toxic or carcinogenic (cancer causing) or both.* Wash your hands

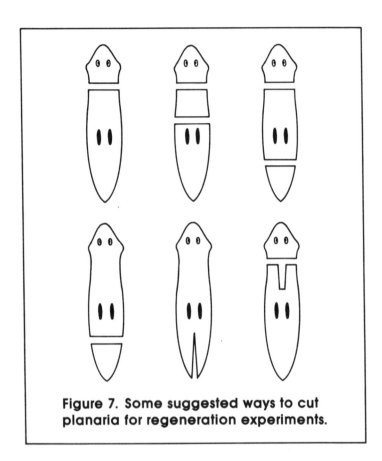

Figure 7. Some suggested ways to cut planaria for regeneration experiments.

with soap and water whenever you take a break from experimentation.

Compare the speed, pattern, and success in regeneration of cutaway portions of planaria by examining the pieces at intervals of 3 or 4 days. Under ideal conditions, most planaria will regenerate the missing portions of their bodies if the pieces are large enough. Do not feed the planaria while regeneration is taking place, but continue to change the water daily, keeping the concentration of silver or chromium ions constant in the experimental dishes.

Take measurements and make drawings each time you examine the planaria.

Did the metal ion have an effect on regeneration at any level tested? If so, is it a level consistent with probable or possible human exposure? Were any parts of the planarian more affected by the metal?

Of course, it is possible to test other agents for teratogenicity using the planarian as a model. Here are some additional suggestions you may wish to explore.

Safety Note: Be sure to work under supervision for the next projects, wear proper lab protection, and wash up after working with chemicals.

- It has been suggested that sleeping under an electric blanket might developmentally harm a human fetus. Devise separate tests of slight temperature elevation and electromagnetic fields on planaria regeneration to investigate this concern. Check your plan with a knowledgeable adult.
- Test other chemical agents, such as fertilizers, crude and motor oils, and detergents. Be sure to work under supervision.
- Test water samples available near your home. You could use runoff water from household gutters, stream runoff near fertilized farm fields, water samples near stockyards (watch out for those bulls and cows!), water from standing puddles. Filter out organic debris, evaporate the water under a fume hood, and redissolve the remaining material in spring water to test on planaria.

RADIATION

Radiation is invisible. You cannot see the sun's ultraviolet rays, but they can still cause sunburn with as little as 20 minutes' exposure. Ultraviolet rays, X-rays, and gamma rays are high-energy forms of radiation capable of causing great damage. Alpha and beta particles, another form of radiation, also cause cellular damage to the lungs and digestive system or to the skin.

Obvious precautions should be taken when investigating radiation. Nevertheless, a number of radiation projects may safely be undertaken.

Safety Note: Experiments involving any form of radiation, including sunlight, should be checked by a qualified science teacher before you begin.

Radon

Radon is a colorless, odorless radioactive gas which seeps into homes and other buildings from the ground around them. Inhaling radon or its decay products can dramatically increase the risk of lung cancer. Not all buildings are affected equally, and most are not affected at all. The geological characteristics of the region determine the likelihood of radon's being present in that area.

- Research the reports of radon concentrations in buildings in your state and make a map. Overlay another map with geological features that help explain the presence of radon in some areas and its absence elsewhere.
- Low-cost radon detectors are offered for sale by several sources, including mail-order businesses. Investigate the efficacy of these devices. Compare with other methods of measurement suggested by geologists, chemists, or environmental engineers, especially those at your state university.
- If radon exposure is a concern in your area, interview contractors to determine what measures, if any, are being taken to control radon levels in the homes and other buildings they construct.

Other Projects with Radiation

- Thorium, an alpha particle emitter, was once used in the bright orange paint found on dinnerware called Fiestaware. This dinnerware is still available at flea markets and secondhand stores. See whether you can locate some and use a Geiger counter to detect the alpha particles.

- Thorium is also found in Coleman lantern mantles. Use the Geiger counter to detect alpha particles. Can they be detected through the packaging in which the mantle is sold? Does the mantle emit more alpha particles before being lit, during use, or after use? Why is thorium present in these mantles?

- Continuous low-level exposure to radiation can be detected with film badges. They are worn by people who work with radiation or radioactive isotopes, particularly in medical labs. Find out how a film badge measures accumulated radiation exposure. Make or obtain your own film badges and place them in various indoor and outdoor locations, especially near large granite buildings. Leave them for several weeks or months. Can you explain any differences in readings you obtain at these locations? Do any readings exceed acceptable background levels?

- Using fruit flies, you can find out whether offspring are affected by a parent's exposure to radiation. Collect your own fruit flies and keep them in cotton-stoppered small glass jars. Observe such features as wing and body shape, body and eye color, and presence or absence of bristles. Consult a biology teacher for instructions on feeding and maintaining the cultures. Ask a hospital, dentist's office, or university to irradiate several jars for you. Continue to keep the flies until they breed and the dark pupae (a larval form) can be seen in the jar. Release the adults and observe the young when they become flies. Have any new characteristics emerged? What percentage, if any, of the young have been affected by the radiation? How did the radiation have its effect? What dose caused an effect?

READ MORE ABOUT IT

"Beverages Intoxicated by Lead in Crystal." *Science News,* January 26, 1991, p. 54.

Biological and Environmental Effects of Low Level Radiation. Vienna: International Atomic Energy Agency, 1976.

Carson, Bonnie L., Harry V. Ellis, and Joy L. McCann. *Toxicology and Biological Monitoring of Metals in Humans,* Chelsea, Mich.: Lewis, 1986.

Meyers, Vera Kolb. *Teratogens: Chemicals Which Cause Birth Defects.* New York: Elsevier, 1988.

"New Lead Rules for Water." *Science News,* May 18, 1991, p. 308.

Paulos, John Allen. *Innumeracy: Mathematical Illiteracy and Its Consequences.* New York: Hill and Wang, 1988.

Paustenbach, Dennis J., ed., *The Risk Assessment of Environmental and Human Health Hazards: A Textbook of Case Studies.* New York: Wiley, 1989.

Upton, Arthur C. *Ionizing Radiation and Health.* Burlington, N.C.: Carolina Biological Supply Company, 1986.

Whitten, Richard H., and William R. Pendergrass. *Carolina Protozoa and Invertebrate Manual.* Burlington, N.C.: Carolina Biological Supply Company, 1980.

7

DRUGS AND THE BODY

Millions of dollars are spent every year educating people about drugs. The reason: drug abuse makes people sick, disrupts their lives, and sometimes kills. What do abused drugs do to the body? How long do they have an effect? Do they have anything to do with overall health? And why, with all the warning messages and information about drugs, do people still abuse them?

Most people abuse drugs because the drugs alter their mood. Almost all abused drugs affect the nervous system in some way. Stimulants such as nicotine, caffeine, and amphetamines speed up nerve cell activity, including that in the brain. They also increase mental alertness and create a sense of well-being. Depressants such as alcohol and sleep-inducing medications slow nerve cell activity.

All of the abused drugs change our thoughts, feelings, and actions because they affect the mind. But they also affect other systems of the body such as circulation and respiration because the nervous system controls other systems as well.

DEPRESSANTS AND *TETRAHYMENA*

Depressants include anesthetics, over-the-counter sleep medications, alcohol, and prescription drugs such as barbiturates. They all dull the responses of the brain and the rest of the central nervous system. Testing for the effects of such drugs in humans is inappropriate and may be dangerous, so microorganisms are suggested for many of these projects.

Although microorganisms lack a nervous system, they are still a good choice for testing the effects of drugs. Their overall responses to certain drugs are often similar to those in animals. A particularly good candidate is one which shows obvious changes in behavior in response to stimulation. *Tetrahymena*, a single-celled organism which moves by way of many tiny hairlike cilia, is an appropriate choice for these kinds of experiments. It is known to react to other environmental stimuli by changing its speed and direction of movement.

How does a depressant like sleep medication affect *Tetrahymena*? Obtain a culture through a biological supply house or collect your own from the water of a clean, quiet pond. Begin your work by becoming familiar with the normal movement of *Tetrahymena*. Place a sample of the culture in the well of a depression slide. Spend some time observing the speed and orientation of the tetrahymena under the microscope.

Next grind up one tablet or capsule of over-the-counter sleep medication in 25 ml of spring water. Add a drop of this solution to the slide.

Do you observe any changes in movement? Is speed affected? Do the organisms change direction more often or less often? Experiment with different dosages.

- Study dose effects, beginning with a small measured amount of the drug and increasing it gradually. What do you observe?

- To get quantitative results, measure the amount of time it takes an individual organism to cross the field of view. Do this for several tetrahymena. Do they all react similarly?
- Attach a video camera to the microscope to collect visual data.
- Compare the actions of several depressant drugs. Try ground-up antihistamines. Add a drop of ethanol to the depression slide in place of the sleep medication. Do these drugs have similar or different effects? Try to use ethanol that has not been denatured.

Safety Note: Ethanol is very flammable. Work away from flames in a well-ventilated room. Know what to do in case of fire.

Legal Guidelines: Work under adult supervision because possession of alcohol by a minor is illegal.

- What physical effects does alcohol have on humans? Are they similar to the effects you saw in the *Tetrahymena* experiment? Are there long-term as well as short-term physical effects? Do some *library* research to examine these questions.

STIMULANTS AND *EUGLENA*

What effects, if any, do stimulants have on microorganisms? *Euglena,* which moves by means of a single or a double whiplike flagellum, is a handy microorganism for this kind of experiment. It uses the green pigment in its chloroplasts along with light to make its own food. You can see a picture of a euglena in Photo 10.

Find out whether caffeine, a commonly used stimulant, affects *Euglena*'s movement and whether it affects this organism's orientation to light.

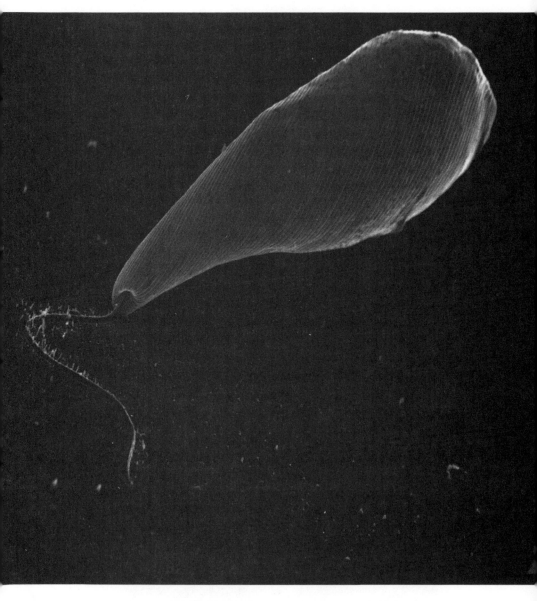

Photo 10. *Euglena* is a good organism to use for investigating the effects of stimulants.

Place a sample of *Euglena* culture (available from a biological supply house) in a depression slide and observe the organism's normal movement. Next shield half of the slide from light. Observe the behavior with respect to light.

Grind up half of a No-Doz tablet or similar over-the-counter caffeine product in 25 milliliters of spring water. Add a drop of this preparation to your *Euglena* sample.

Do you see any changes in *Euglena*'s movement or orientation to light? If there is a change, what can you conclude about the effects of caffeine on speed, light preference, or light avoidance? Are changes, if any, immediate? How long do they last?

- Look at the dose effect by adding an even smaller amount of caffeine at first and then gradually increasing the amount.
- Try other stimulants, such as over-the-counter diet drugs and nicotine. To obtain nicotine construct a smoking machine like the one pictures in Figure 8. Connect the rubber tubing to an aspirator and burn a complete pack of cigarettes to obtain enough nicotine for your experiment. Be sure to do the nicotine extraction in a fume hood. Do all the stimulants which you have tested have the same or similar effects on *Euglena* behavior?

Legal Guidelines: Do all nicotine experiments under supervision if you do not meet the age requirement for tobacco possession in your state.

- Do stimulants affect other microorganisms in a similar way? Experiment with caffeine, nicotine, or diet drugs on *Tetrahymena, Paramecium,* rotifers, or other microorganisms you can collect.

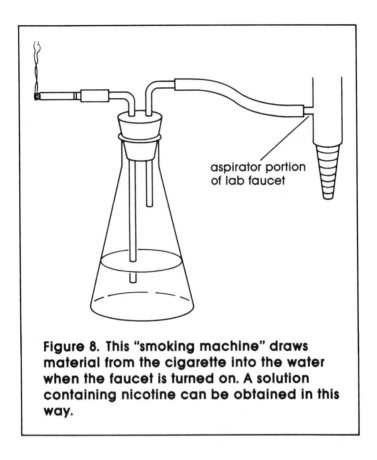

Figure 8. This "smoking machine" draws material from the cigarette into the water when the faucet is turned on. A solution containing nicotine can be obtained in this way.

Here are some additional *Euglena* experiments, using both over-the-counter stimulants and depressants, to try:

- Design a project to find out whether the color preferences of *Euglena* help ensure its survival. Do any of the stimulants, such as caffeine or nicotine, or depressants, such as sleep medications, influence *Euglena*'s color preference? If so, how might it affect the organism's survival?

- Administer a stimulant and then a depressant in series to *Euglena.* Does combining these two kinds of drugs have any effect? Does the order of administration change the outcome?

THE EFFECTS OF NICOTINE ON *DAPHNIA*

Smokers often say that smoking gives them a sense of well-being. It is the nicotine in cigarette smoke which affects the nervous system in this way, but does nicotine have effects elsewhere in the body? You can study the effects of nicotine in a tiny crustacean, the water flea *Daphnia* (see Photo 11).

Two characteristics make *Daphnia* a good choice for this type of project. First, a daphnia has organ systems roughly analogous to human systems, for example, a circulatory system, controlled in part by the nervous system. Second, the internal organs, including the heart, are visible through its transparent body, as shown in Figure 9.

How do you think nicotine will affect *Daphnia*'s beating heart?

Obtain a *Daphnia* culture from a biological supply company. Place a single daphnia on a slide, add only enough water to keep it alive, and do *not* use a coverslip or you will crush it.

Observe the beating heart under low power of the microscope. Practice timing the heartbeat with a stopwatch. Record these heart rate measurements. They will be the control for your experiment.

Prepare a water solution of nicotine with the smoking machine described earlier. Be sure to work under adult supervision and use a fume hood.

Next add a drop of your nicotine solution to the slide and observe for several minutes. Is there an effect? If so, how soon do you notice a change? How long does it last? What do you conclude?

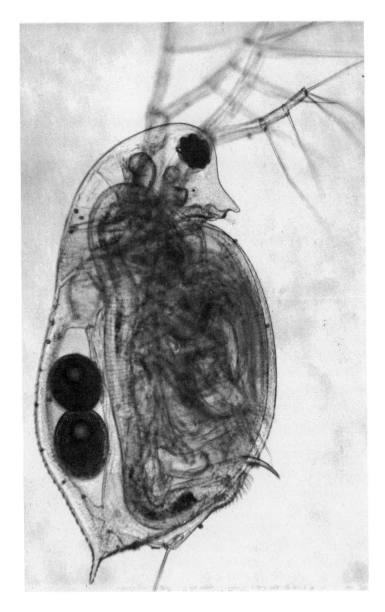

Photo 11. *Daphnia,* the water flea, can serve as the subject of an investigation of the effects of nicotine on heart action.

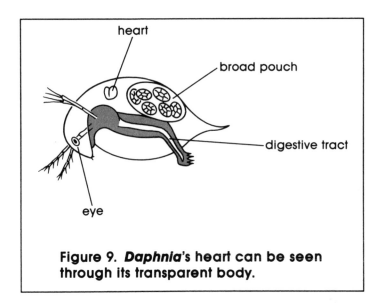

Figure 9. *Daphnia*'s heart can be seen through its transparent body.

Here are some related projects:

- If there is a change in heart rate, check other daphnia to see whether they all react similarly.
- Do other stimulants affect *Daphnia*'s heart rate? Crush diet medications or tablets containing caffeine in distilled water to make dilute solutions. Test these with the daphnia. Use only dilute solutions to minimize the possibility of killing your test organisms.
- Transfer a small amount of asthma inhalant, another over-the-counter drug, to a daphnia on the slide. Compare the heart rate to the normal *Daphnia* heart rate. Do you see an effect? What can you conclude about asthma inhalant from your observations?
- Do depressants affect *Daphnia*'s heart rate? With *adult supervision*, administer sleeping medica-

tions, ethanol, or antihistamines to *Daphnia.* What effect, if any, do you think each of these will have? Will any of these act in the same way? What can you conclude about the effects of depressants on heart rate?

• Investigate the use of other *invertebrates* to study the effects of nicotine or caffeine on a system other than the nervous system. What characteristics are desirable in an experimental organism? *Never use vertebrates for this experiment.*

DAPHNIA AND MULTIPLE-DRUG EFFECTS

You may be aware that coma and death can result when people mix alcohol and other depressants. Choose two or more over-the-counter-drugs whose effects on *Daphnia* you have already observed. Develop a hypothesis about the combined effect of the two drugs, expressed in quantitative terms. Administer the combination to a daphnia and observe. Repeat with other daphnia.

• Try different combinations, that is, two stimulants, two depressants, and a stimulant/depressant combination. Are combination effects readily predictable from evidence of separate effects?

Here are four questions to investigate:

• Are there warnings about combination drug use found on over-the-counter medicines? Can you identify any dangerous combinations? Are stimulants and depressants frequently mentioned?
• Does long-term damage to organisms occur after repeated applications with the same drug? Form a hypothesis concerning this and then carry out a test with *Daphnia.*
• Is the accidental and dangerous combination

of over-the-counter drugs a serious problem? Interview pharmacists, physicians, and emergency
room staff to find out.

• Do *Daphnia* develop "tolerance" to stimulants
  or depressants?

CAFFEINE AND PULSE RATE

Trained athletes often have resting pulse rates that reach
down into the 40s and 30s. This is well below the 60s to
80s of normal adults. It is generally agreed that, as a result
of aerobic training, the athlete's heart has become more
efficient. Although this low resting heart rate is considered
healthy, whenever drugs raise or lower anyone's heart rate,
they can pose a potentially serious health risk.

In this project, you will be working with human subjects, so you need to apply the following guidelines.

Guidelines for Working with Human Subjects: check
Chapter 1 for special guidelines to follow when working
with human subjects. Volunteers for these projects *must* be
medically screened. *Do not* use volunteers who have respiratory or cardiac problems. Ask a science teacher and
a health professional to check your experimental plan before proceeding.

Legal and Ethical Guidelines: subjects for experiments involving tobacco must be of legal age for tobacco use and already smokers. Subjects for experiments
involving caffeine must already be caffeine consumers.

The question you will be investigating is, Does caffeine
affect heart rate in humans? To find out, select a suffi-
ciently large group of volunteer subjects who regularly
consume caffeinated cola beverages. Ask them to ab-
stain from caffeinated drinks and any other stimulants, such
as nicotine, on the day of the test.

Divide your volunteers into two groups and determine
the resting pulse of each individual. Give one group a
predetermined, reasonable amount—for example, one or

two 12-ounce cans—of a caffeinated cola drink and the other group the same amount of an uncaffeinated cola of the same brand. (Do you think you should tell each group which cola it is getting?) Then take pulses at several timed intervals after cola consumption.

Does caffeine affect pulse rate? If so, when is the effect first noticeable? How long does it last?

- The harder you exercise, the faster your heart beats. Vigorous exercise should take place in the *target heart rate* range and never reach *maximum heart rate*. Find out what target and maximum heart rates are and how they are calculated. Then design an experiment to test the effect, if any, of caffeine on target heart rate in exercise.

  Does caffeine affect the activity level required to reach target heart rate? Is it wise to consume caffeinated beverages before vigorous exercise such as running in a cross-country meet or playing basketball?
- Find out why some individuals with some circulatory conditions are advised to limit coffee intake.
- Smoking a cigarette is a habit that often accompanies coffee drinking. Predict the combined effects of caffeine and nicotine on pulse rate. You might extrapolate from *Daphnia* results obtained from experiments in the previous section.

CAFFEINE AND THE NERVOUS SYSTEM

Stimulants and depressants affect the way messages are transmitted by the brain and nerves. Could a stimulant such as caffeine affect the taste of food, the sensitivity of your skin, or your ability to concentrate? Are any such changes in a person's response to his or her surroundings measurable?

Guidelines for Working with Human Subjects: See the guidelines in the previous section.

Here are some projects involving changes in human nerve cell responses that may be caused by caffeine:

- Use a box of straight pins to test skin sensitivity on various regions of the body. Tape the pins in parallel pairs, so that one pair has its points 1 millimeter apart, another, 2 mm apart, etc. Test the skin on the fingertips, back of the hand, back of the neck, arm, and lip by touching the pin pairs to the skin while the subject is not looking. Ask the subject whether the pins feel like one point or two points. When a subject is able to identify two very close pin points, the sensitivity is considered high. Does caffeine, in amounts normally consumed, affect the sensitivity of these skin receptors? What are the consequences of either heightened or depressed skin sensitivity to an individual?

- Coffee and tea drinkers claim that their beverages help them get started in the morning, focus their attention, and accomplish tasks more effectively. Carry out an experiment to test this claim. Think of a task which is measurable and which requires concentration or dexterity. Ask volunteers to perform the task. Does caffeine have an effect on how quickly or how accurately the task can be completed? If so, how much caffeine is required to make a difference? Be careful of the "training effect" when you ask the same person to repeat the same or similar tasks several times.

- Does caffeine affect reaction time? Many driver education programs have simple machines which measure reaction time. You may be able to use the one at your school or a device of your own construction to answer this question.

READ MORE ABOUT IT

Gilbert, Richard J. *Caffeine: The Most Popular Stimulant.* New York: Chelsea House, 1986.

Hyde, Margaret O. *Know about Drugs.* New York: Walker, 1990.

Julien, Robert M. *A Primer of Drug Action.* San Francisco: W. H. Freeman, 1985.

Lindblad, Richard. *Drug Abuse.* Burlington, N.C.: Carolina Biological Supply Company, 1984.

Perry, Robert L. *Focus on Nicotine and Caffeine.* Frederick, Md.: Twenty-first Century Books, 1990.

Schlaadt, Richard G., and Peter T. Shannon. *Drugs.* Englewood Cliffs, N.J.: Prentice, Hall, 1990.

Weil, Andrew and Winifred Rosen. *Chocolate to Morphine: Understanding Mind-Active Drugs.* Boston: Houghton Mifflin, 1983.

Weiss, Ann. *Over-the-Counter Drugs.* New York: Franklin Watts, 1984.

8

GENETICS

The relatives of the new baby crowd around the hospital's nursery window. One observer exclaims, "She has her mother's eyes, but that's definitely *his* nose." Such experiences in families confirm a basic rule of genetics, that both parents contribute to the inherited characteristics of their children.

This rule derives from investigations made in the nineteenth century by Gregor Mendel. But Mendel didn't study human noses to make his discoveries. He studied peas, which, in reproducing quickly and abundantly, provide a great deal of information in a short period.

Important as they were, Mendel's discoveries only scratched the surface in our understanding of heredity. The tools and discoveries of the twentieth century have increasingly thrown light on the subject, allowing details and surprises about heredity to emerge. During the years you are in high school, the knowledge we have about human heredity will more than double. Yet our understanding of the genetic contribution to such obvious human traits as eye and skin color, as well as less definable traits such as intelligence, is far from complete.

SINGLE-GENE INHERITANCE

Among Mendel's contributions was the description of a type of inheritance based upon a single factor inherited from each parent. Today we call such a factor an allele, an alternative form of a gene. In the simplest form of inheritance, the two alleles, one from each parent, control a single inherited characteristic such as hairline. One parent may donate an allele for a rounded hairline, *w*, and the other parent may donate an allele for a *v*-shaped hairline called widow's peak, *W*. The allele for the widow's peak form of the trait has been given the capital letter W because it has been found to express *dominance* over the other allele w, which is *recessive*. When the two alleles combine in one individual, that person will have widow's peak. Figure 10 illustrates the difference between widow's peak and rounded hairline.

Thousands of human characteristics are reported to be expressed as a single pair of alleles. Such inheritance patterns are found by studying genetic family trees, called pedigrees by geneticists. These trees indicate the mechanism of inheritance for a particular characteristic within a family. Refer to Figure 11 for an example of a family pedigree for the trait polydactyly, having extra fingers and toes. Figure 12 is the key to the symbols in Figure 11.

Among human characteristics thought to be controlled by a single gene are several whose mode of inheritance is uncertain. One which has been presented in science textbooks for years as a dominant trait is the ability to roll the tongue (Photos 12 and 13). However, there is some evidence that the ability to roll your tongue is not strictly an inherited trait.

Is the ability to roll the tongue a single-gene trait with a dominant inheritance, as most texts say, or can tongue

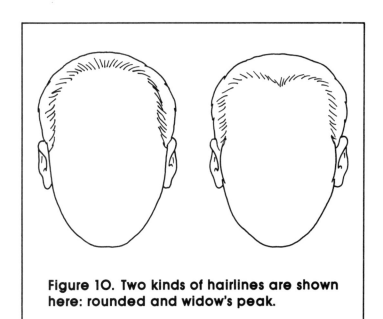

Figure 10. Two kinds of hairlines are shown here: rounded and widow's peak.

rolling be learned? A strong study could be made, using three lines of evidence.

First, collect information on tongue rolling from as many families as possible. Wherever you can, collect information from at least three generations of each family. Construct a pedigree for each family, using the key to symbols found in Figure 12. Shade in the individuals who can roll their tongues.

Next analyze the pedigrees. If you find cases of adult tongue rollers who have both tongue-rolling and non-tongue-rolling children, what does this tell you? If one parent is a roller and the other a nonroller, does this help you decide on the pattern of inheritance? Do any nonrolling parents have children who can roll their tongues? Does this help you?

If you have time, you may wish to pursue a second

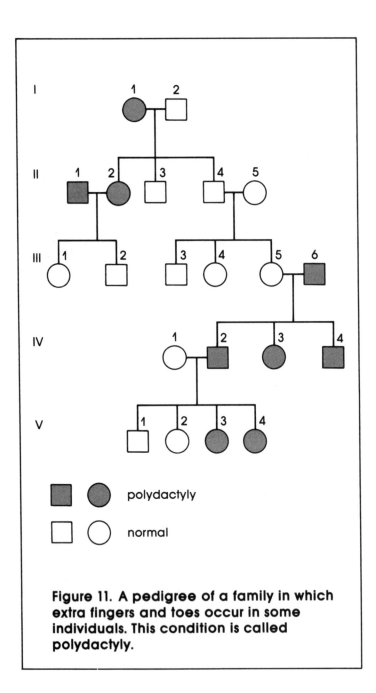

Figure 11. A pedigree of a family in which extra fingers and toes occur in some individuals. This condition is called polydactyly.

Photos 12 and 13. Some people can roll their tongue, some cannot. Conduct a research project to accept or reject the widely held belief that tongue rolling is an inherited trait.

line of inquiry to find out whether tongue rolling is purely inherited or whether it also can be learned. Attempt to teach nonrollers to roll their tongues. Allow several months for practice and recruit several volunteers who will practice regularly. What do you find out?

The third line of pursuit will be the most difficult. To complete it, you will need to find sets of identical twins, or data about identical twins. The University of Minnesota in Minneapolis and Medical College of Virginia in Richmond are two sites with ongoing identical-twin studies. Survey them or review the data to confirm or refute tongue rolling as a simple dominant trait. If you find identical twins (people with exactly the same genes) whose tongue-rolling capabilities differ, what have you learned?

Here are some additional investigations of human traits:

- The ACHOO syndrome has been tentatively reported to be a dominant single-gene trait. It consists of uncontrolled sneezing (at least two sneezes) that occurs when a person goes from darkness into bright sunlight. Two studies, by different individuals in different locations, have found ACHOO syndrome in 23 percent and 36 percent of a surveyed population. What percentage of your school population exhibits ACHOO syndrome? Do pedigree analyses support the dominant inheritance proposed for this trait?
- Clasp your hands, intertwining the fingers. Notice which thumb is on top. Now try clasping the hands with the other thumb on top. For most people one way feels more comfortable. Use pedigree or twin studies or discover whether a single dominant gene controls this behavior.
- Fold your arms in front of your body. Notice which arm is top. As with hand clasping, in arm folding there appears to be a preference for one arm or the other on top. Does a single dominant gene control this behavior?

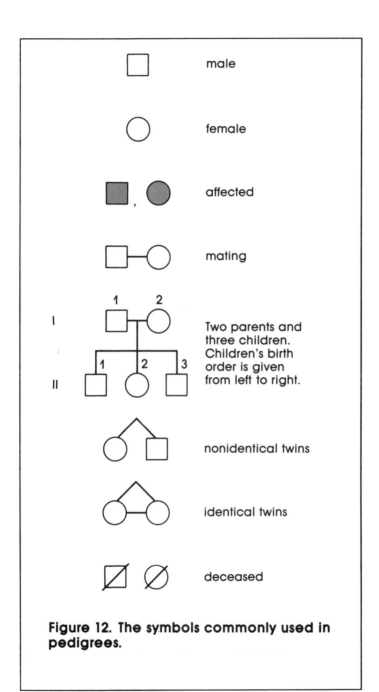

Figure 12. The symbols commonly used in pedigrees.

- There are reported ethnic differences in the relative lengths of the big toe and second toe. Studies of three groups—U.S. whites, Japanese Ainu, and Swedish white males—have shown that 24, 90, and 3 percent, respectively, of the individuals in these three groups have a longer second toe. Are there significant ethnic differences in this characteristic among the students of your school? Are males significantly different from females with respect to this trait? How could gender or ethnic differences have come about?
- The longer second toe is felt to be inherited as a single dominant trait, but one researcher claims that this characteristic has "reduced penetrance." Find out what *penetrance* means in genetics and investigate this claim with pedigree data you collect.
- Are dimpled (cleft) chins the result of a dominant allele? Most studies of cleft chin inheritance are quite old and limited to twenty-five or fewer cases. Design and carry out a more extensive study of the question.
- The ability to taste phenylthiocarbamide (PTC) is conferred by a single dominant allele. PTC is chemically related to several compounds found in vegetables of the mustard family—cabbage, cauliflower, broccoli, and brussels sprouts. PTC taste papers can be ordered from biological supply companies. Are there tasting or preference differences regarding these vegetables between PTC tasters and nontasters?

DNA: THE HEREDITY MATERIAL

Most human traits are inherited in a more complex manner than the single-gene mechanism described by Mendel. Sometimes several genes working together contribute to the expression of a trait. Such is the case with hair color.

In other cases, the environment acts on the expression of a gene or genes to produce an even greater variety in that trait. For example, sunlight can affect skin color, and nutrition can influence height.

Despite different hereditary mechanisms, all the genes of all organisms have one thing in common: they are primarily composed of a material called deoxyribonucleic acid, or DNA. That DNA is the "stuff" of life was discovered in the middle of this century. A series of important discoveries about DNA, its structure, how it replicates, and how it works, followed. Three scientists—James Watson, Francis Crick, and Maurice Wilkins—received Nobel prizes for figuring out the structure of DNA, highlighting the central importance of the substance. Today every student of biology hears and reads of DNA.

Which tissues in which organisms contain the greatest amounts of DNA? When considering the tissues of a large organism, where would you expect to find a high concentration of DNA? Begin with a hypothesis concerning the most likely tissues to contain large amounts of DNA. You can then test your hypothesis by extracting and quantifying DNA from various tissues. Slaughterhouses, also known as abattoirs, and veterinary clinics can often provide the animal tissues you need. Plant tissues are easier to obtain.

To obtain DNA from tissues, you will need to perform an extraction procedure. Chemicals and equipment required are generally available in a chemistry lab. Ask a chemistry teacher or other professional for assistance in ordering chemicals and learning techniques, especially if you have never attempted this type of procedure in the past.

Safety Notes: Work under the supervision of a qualified science teacher. Wear safety goggles, protective gloves, and a lab apron. Isopropanol (isopropyl alcohol) is very flammable; know where the fire safety equipment is and how to use it. Sodium dodecyl sulfate may irritate your skin. Wash your hands after working with chemicals.

You will need to keep all reagents and materials as

cold as possible while extracting. Find the exact mass of 2 to 3 g of animal tissue. To the tissue, add 1 ml of cold buffer (0.87 g NaCl + 10 ml 0.1 M EDTA + 90 ml distilled water), and 2 to 3 drops of sodium dodecyl sulfate. Then grind the tissue.

Add 6 ml 2 M NaCl to your extract in a centrifuge tube and shake vigorously for 2 to 3 minutes. Centrifuge the tube at high speed. Be sure to balance the centrifuge before operating it. Pour off the liquid on top and save. This is the layer containing the DNA. Discard the material in the bottom of the tube.

Slowly add 10 ml of cold isopropanol down the sides of the test tube. The DNA will form at the boundary between the two liquids. Working under supervision, melt the tip of a Pasteur pipette in the flame of a burner and tilt it until it bends into the shape of a hook. Dip the *cooled* hook to the interface of the liquids and twirl it. DNA fibers should collect on the hook. Dip the hook to the bottom of the tube from time to time to collect as much DNA as possible.

Examine the material on the pipette. This is DNA. To prepare it for further examination, twirl the pipette in a test tube containing 2 ml distilled water. After 20 minutes, look at a drop or two with the microscope. To quantify the amount you extracted, you will need to redissolve the DNA. To accomplish this, stopper the tube and let it stand until the next day.

A Spectronic 20 or other colorimeter can be used to quantify the DNA in a sample. Set the absorbance at 500 nanometers (nm). Use distilled water as your blank. Establish a baseline reading. Then make serial dilutions from your 2–ml redissolved DNA sample in order to find a proper concentration for your readings. For more detailed discussion of this procedure, consult a genetics lab manual, such as the one listed at the end of the chapter.

Repeat the entire extraction procedure for each tissue examined. Compare absorbances to determine the tissues with the highest concentration of DNA. The higher the absorbance, the greater the amount of DNA present. Did you find DNA in all the tissues? Were the amounts generally consistent or did they differ? Were there any surprises? How did your hypothesis fare?

MUTATIONS AND DNA

Only a few years after DNA was shown to be the heredity material, George Beadle, a researcher at Stanford University, figured out a way to learn how it works. "One ought to be able to discover what genes do by making them defective," he said in his famous paper "The Genes of Men and Molds." And that is exactly what he did. He made the DNA of a mold defective by subjecting it to X-rays. Each piece of DNA so treated could no longer specify the production of a protein.

Changes in the DNA such as the ones caused by X-rays are called mutations. Most mutations are harmful because they result in a changed protein. The protein usually can't do its job if it has been altered in this way. What other agents besides X-rays cause mutations?

One way to explore this question is to use the bacterium *E. coli,* whose DNA, like that of other organisms, is subject to a low level of mutation at all times. Sometimes this natural mutation causes the bacterium to be resistant to antibiotics as noted in earlier chapters of this book. Such a change is easily detected when *E. coli* will grow in the presence of an antibiotic. But will this rate of mutation increase if *E. coli* are exposed to the ultraviolet light from a germicidal lamp or a tanning lamp?

Safety Note: Avoid directly exposing yourself to the light of a germicidal or tanning lamp. Ultraviolet (UV) light can cause skin cancer and damage to eyes. Use UV-blocking sunglasses and do not look directly at the bulb while the lamp is on. At the conclusion of the experiments described below, autoclave or pressure cook all used petri plates before disposing in marked biohazard bags.

Before beginning this investigation, familiarize yourself

with basic sterile microbiological techniques. The materials you will need are available in many high school biology labs. *E. coli* culture, agar, and antibiotics may be ordered through biological supply houses.

Establish a culture of *E. coli* in a 500-ml flask of nutrient agar. Incubate at 37° C for 24 hours. (A heating pad set on low substitutes well for an incubator.) Have available several agar plates with and without a selected antibiotic in the agar itself. Possible antibiotics to use include penicillin, kanamycin, tetracycline, and chloramphenicol.

Transfer 10 ml of the culture into each of two test tubes. Return one tube to the incubator. Pour the contents of the second tube into an empty sterile petri dish. Expose this dish to the germicidal lamp for 30 seconds or to the tanning lamp for the recommended tanning time. Return the plate to the incubator.

After 30 minutes' incubation, use a sterile spreader (refer to Chapter 5), to spread 0.5 ml of the control culture onto control plates containing no antibiotic and plates containing antibiotic in the medium. Do the same for your ultraviolet-light-treated sample. Then dilute each culture to one-tenth the concentration and repeat the spreading procedure on new plates. Use care in labeling throughout. Let the plates stand for 5 minutes, then incubate for 24 hours at 37° C.

Count colonies on control plates containing antibiotics to calculate a background mutation rate. Count colonies on ultraviolet exposed plates and compare. If you get *no* growth on the light-exposed plates, repeat the experiment with a shorter exposure to the light. Did the ultraviolet light increase the mutation rate? How do you know? If increased mutation did occur, does this say anything about the advisability of exposure to ultraviolet light?

Here are some additional projects about mutation which you may wish to explore:

- Are some portions of the DNA more susceptible to mutation than others? Repeat the experiment several times, using a different antibiotic in the agar each time. For valid comparison, be sure to keep all other factors unchanged.

- Because of frequent reporting about cancer and environmental exposures, some people think that almost everything causes this disease. Real cancer-causing agents, or carcinogens, are usually also mutagens. Design an experiment using seeds, root tips, bacteria, or lily anthers to test your own hypothesis about the mutagenicity of various substances.
- Can mutagenesis be reduced by substances promoted as cancer protectors, such as beta-carotene?
- The use of the preservative formaldehyde in the science classroom has been significantly reduced for safety reasons. Is the amount of this product found in cosmetics and fabric finishes mutagenic?
- Investigate the Ames test for mutagencity. What are the advantages of this test? Can it or some variation of it be performed by students?

HEREDITY AND THE ENVIRONMENT

One of the questions most frequently asked about human traits is, How much is inherited and how much is affected by the environment? For some traits, it's one way or the other, but for most human traits, the answer is probably that some of each are at work. How *much* of each is the hard part to determine.

One human trait, handedness, has been the subject of frequent debate over genetic and environmental influences for years. One hypothesis suggests that two alleles for right-handedness mean a person will always be right-handed. Likewise two alleles for left-handedness will produce a lefty. But an individual with one allele for each can show a preference for either left- or right-handedness.

- Is this hypothesis supported by extensive pedigree or identical twin data?

    Guidelines Working with Human Subjects. All investigations using people require you to obtain their permission. See Chapter 1.

- Is handedness as a trait limited to humans? Is there handedness among cats or dogs? To answer these questions, obtain permission from pet owners in your neighborhood to test their cats or dogs for handedness. First get to know the pets with whom you will be working so their cooperation will be more likely. Supplying treats is a great way to be accepted.

    Construct some form of barrier with a hole in it. It should be lightweight and portable so it can be taken to the pet's home. Place the barrier between you and the pet. Offer a treat to the pet from just behind the hole. Design the hole so that the pet must reach through the hole to get the treat. Note which paw the animal uses. If the pet cannot retrieve the treat, reward it for trying. Return to test the same group of pets often so patterns, if they exist, may be observed. Obviously the greater the number of cats or dogs you test, the greater your chances of learning something significant.

    If you know an animal breeder, it may be possible for you to test litters of dogs or cats. This would allow you to establish a handedness pedigree for mother and offspring, if you find that these animals do exhibit handedness preference.

- Each person's fingerprints are unique, but the pattern and total number of ridges on the ten fingers are considered to be under the influence of as many as seven genes and the environment. Find out how fingerprints are classified and how total ridge counts (TRCs) are taken. Then design a study of fingerprint patterns and total ridge count in families or among identical twins. How can similarities and differences be accounted for?

STUDENT RESEARCH IN BIOTECHNOLOGY

Biotechnology is a descriptive term applied to areas of research and the specific laboratory techniques that use organisms to make products. Today, thousands of scientists, in both the public and private sectors, are engaged in biotechnology research, most of it oriented toward genetics. Medical biotechnology includes recombinant DNA work, monoclonal antibody experiments, vaccine development, prenatal diagnosis, DNA fingerprinting, and human genome mapping. Since these fields are advancing so rapidly, periodicals are a better source of the latest information than books. And since there's a good chance a biotech facility—either a business or a university lab—is operating near you, if you are seriously interested in this field, then visiting one of these facilities, and perhaps getting to know some scientists working there, is a good idea.

There are many restrictions on the types of biomedical research that can be done in certain facilities, such as schools, or by high school students outside a research setting. More and more high schools now have the equipment and teachers trained to do recombinant DNA experiments involving bacteria, but the National Institutes of Health does limit work to prokaryotic (bacterial) genes. If you are enthusiastic about this kind of work, here is a plan you may want to follow.

1. Do some background *reading* in several areas which interest you. Learn a few basic procedures and some of the ideas involved.

2. Contact state universities or colleges, biotechnology firms, state and federal government labs, and crime labs. Consult adults for suggestions for whom to contact.
3. Convey your enthusiasm, knowledge, and willingness to undertake a project or to help with ongoing research.
4. Be realistic about the kind of work you may do.
5. Find out about summer research programs for students.
6. Follow up on all leads.
7. *Stick* with it. Who knows where this may lead? An interesting discovery, new skills, knowledge about a career? And by following through, you'll be helping the next person who'd like to try.

READ MORE ABOUT IT

Boulnois, Graham J. *Gene Cloning and Analysis: A Laboratory Guide*. Oxford: Blackwell, 1987.

Gardner, Elton J., Thomas R. Mertens, and Robert L. Hammersmith. *Genetics Laboratory Investigations*. Minneapolis, Burgess, 1985.

Gardner, R.J.M., and Grant R. Sutherland. *Chromosome Abnormalities and Genetic Counseling*. New York: Oxford University Press, 1989.

Gerbi, Susan A. *From Genes to Proteins*. Burlington, N.C.: Carolina Biological Supply Company, 1987.

Gross, Cynthia. *The New Biotechnology*. Minneapolis: Lerner, 1988.

Lee, Essie. *A Matter of Life and Technology*. New York: Julian Messner. 1986.

Maxson, Linda R., and Charles H. Daugherty. *Genetics: A Human Perspective*. Dubuque, Iowa: William C. Brown, 1992.

McKusick, Victor. *Mendelian Inheritance in Man*. 8th ed. Baltimore: Johns Hopkins University Press, 1990.

Micklos, David, and Greg Freyer. *DNA Science: A First Lab Course in Recombinant DNA Technology*. Cold Spring Harbor, N.Y.: Cold Spring Harbor Laboratory, 1990.

Milunsky, Aubrey. *Choices, Not Chances.* Boston: Little, Brown, 1989.

Pierce, Benjamin. *Family Genetic Sourcebook.* New York: John Wiley, 1990.

Stockton, William. *Altered Destinies: Lives Changed by Genetic Flaws.* Garden City, N.Y.: Doubleday, 1979.

Thompson, Margaret W., Roderick R. McInnes, and Huntington F. Willard. *Genetics in Medicine.* Philadelphia: Saunders, 1991.

Wingerson, Lois. *Mapping Our Genes.* New York: Dutton, 1990.

APPENDIX: SCIENTIFIC SUPPLY COMPANIES

The companies listed here provide specimens, labware, chemicals, and other supplies to schools. Most will sell items to individuals if a check or other form of payment accompanies the order. Consult your teacher to look at catalogs and compare terms.

The following list represents some of the larger companies. After each listing the following code appears to inform you of the nature of the products:

S = Living specimens
C = Chemicals
L = Labware
O = Other supplies

Carolina Biological
   Supply Co.
200 York Rd.
Burlington, N.C. 27215
(919) 584–0381
(800) 334–5551
S. C. L. O

Chem Scientific Co,
11222 Melrose Ave.
Franklin Park, IL
   60131
(708) 451–0150
(800) 262–3626
C. L. O

ABOUT THE AUTHOR

Karen E. O'Neil teaches biology and other science courses at Annie Wright School in Tacoma, Washington. She received her B.A. and M.S. in zoology from the University of Maine at Orono. For a short time, she worked as a research assistant at Harvard Medical School. During the past twenty years, she has taught in public and independent schools in five states, where she has encouraged students' curiosity and competition in science fairs. Ms. O'Neil has written articles and contributed to other publications. This is her first book.

PROJECTS FOR YOUNG SCIENTISTS

EACH BOOK IN THE PROJECTS FOR YOUNG SCIENTISTS SERIES TELLS YOU WHAT A SCIENCE PROJECT IS AND HOW TO DO ONE, DISCUSSES SCIENCE FAIRS, AND PRESENTS NUMEROUS IDEAS FOR SCIENCE PROJECTS SUITABLE FOR CLASSROOM ASSIGNMENTS OR SCIENCE FAIRS.

AVAILABLE IN PAPERBACK:

BIOLOGY

BOTANY

ECOLOGY

ENERGY

ENGINEERING

GENETICS

GEOLOGY

THE HEART AND CIRCULATORY SYSTEM

MATH

NATURE

SPACE SCIENCE

08-ALT-631

$6.95 U.S.
$9.99 CANADA